CHAOS

A Tool Kit of Dynamics Activities

ROBERT L. DEVANEY

JONATHAN CHOATE

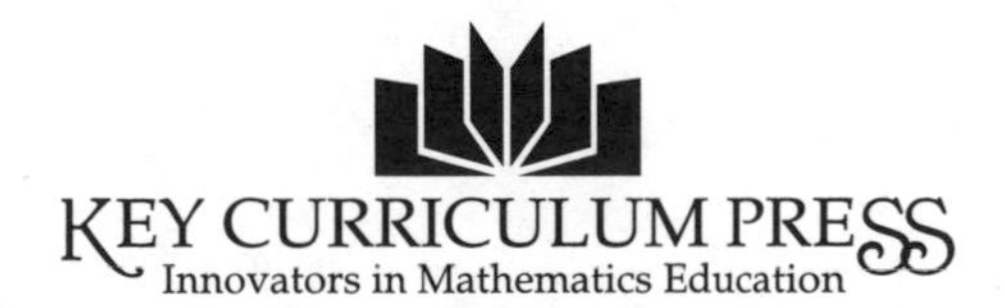

Editor	Casey FitzSimons
Project Administrator	James A. Browne
Production Editors	Jason Luz, Jennifer Strada
Copy Editor	Margaret Moore
Production and Manufacturing Manager	Diana Jean Parks
Production Coordinator	Ann Rothenbuhler
Text and Cover Designer	Kirk Mills
Compositor	Ann Rothenbuhler
Technical Artists	Ben Turner Graphics
Art and Design Coordinator	Caroline Ayres
Prepress	TSI Graphics
Printer	Versa Press, Inc.
Publisher	Steven Rasmussen

This material is based upon work supported by the National Science Foundation under award number ESI-9255724. Any opinions, findings, and conclusions or recommendations expressed in this publication are those of the authors and do not necessarily reflect the views of the National Science Foundation.

Key Curriculum Press
1150 65th Street
Emeryville, CA 94608
510-595-7000
editorial@keypress.com
http://www.keypress.com

Printed in the United States of America 10 9 8 7 6 5 4 3 2 03 02 ISBN 1-55953-356-0

CONTENTS

LETTER FROM THE AUTHORS v

INTRODUCTION: WHY CHAOS? ix

Lesson 1 ▹ Linear Iteration

MS	A1	G	A2	P/C
◑	◉	●	◉	◉

TEACHER NOTES 1
EXPLANATION 5
INVESTIGATIONS 10
FURTHER EXPLORATION 17

Lesson 2 ▹ Types of Fixed Points

MS	A1	G	A2	P/C
◔	●	●	◉	◉

TEACHER NOTES 19
EXPLANATION 22
INVESTIGATIONS 26
FURTHER EXPLORATION 37

Lesson 3 ▹ Graphical Iteration

MS	A1	G	A2	P/C
◔	◉	●	◉	◉

TEACHER NOTES 39
EXPLANATION 42
INVESTIGATIONS 50
FURTHER EXPLORATION 63

Lesson 4 ▹ Measuring Population Growth

A1	G	A2	P/C
●	◑	◉	◉

TEACHER NOTES 65
EXPLANATION 68
INVESTIGATIONS 73
FURTHER EXPLORATION 80

Curriculum Correlation Key

◔ Small portion of lesson is relevant to indicated stage of curriculum.

◑ About half of lesson is relevant.

● Entire lesson is relevant.

◉ Lesson is particularly relevant and could replace a traditional lesson.

MS Middle School

A1 Algebra

G Geometry

A2 Algebra 2

P/C Precalculus or Calculus

Lesson 5 ▹ Nonlinear Iteration

A1 ● G ◑ A2 ◉ P/C ◉

Teacher Notes 81
Explanation 85
Investigations 92
Further Exploration 104

Lesson 6 ▹ Chaos

A1 ● G ◑ A2 ◉ P/C ◉

Teacher Notes 107
Explanation 111
Investigations 121
Further Exploration 127

Lesson 7 ▹ The Butterfly Effect

A1 ● G ◑ A2 ◉ P/C ◉

Teacher Notes 129
Explanation 133
Investigations 139
Further Exploration 143

Lesson 8 ▹ Cycles and Nonlinear Iteration

A1 ◑ G ◑ A2 ◉ P/C ◉

Teacher Notes 145
Explanation 150
Investigations 162
Further Exploration 166

Lesson 9 ▹ The Orbit Diagram

A1 ◑ G ● A2 ◉ P/C ◉

Teacher Notes 169
Explanation 172
Investigations 182
Further Exploration 185

Lesson 10 ▹ A Quadratic Expedition

A1 ◑ G ◑ A2 ● P/C ◉

Teacher Notes 187
Explanation 189
Investigations 191

Answers 203

Letter from the Authors

Dear Educator,

Welcome to *Chaos: A Tool Kit of Dynamics Activities,* the third book in our collection of workbooks designed to introduce students to the marvelous new area of mathematics called dynamical systems. From the first day we began to expose our students to these contemporary ideas, we have been committed to helping others spark interest in their students by exposing them to the beauty and mystery of these rich mathematical topics. In assembling this book, we have attempted to provide you with a blend of explanations, investigations, and teacher information to facilitate your journey into this chaotic world!

Curriculum Links

Although this book can be used as a stand-alone supplementary unit in many mathematics courses, its real strength lies in integrating it within the existing secondary school mathematics curriculum by treating it as a contemporary strand in mathematics. Many of the topics covered in this book mesh very nicely with topics in algebra courses, and the use of technology is an essential feature of this topic. For example, finding fixed points involves solving linear or nonlinear equations. Graphical iteration demands knowledge of not only how to graph a function, but also what the graph means explicitly. Iterating the simples of nonlinear functions demands use of technology, employing the use of "list" features on calculators or the recursive features of spreadsheets.

The Contents pages contain icons that help you identify how specific lessons correlate with different points in the traditional curriculum.

Lesson Contents

At the beginning of each lesson you will find a set of **Teacher Notes.** These notes begin with a short *Overview* of the lesson, which briefly describes the main focus of the lesson. Next we discuss the *Mathematical Prerequisites* and *Mathematical Connections* for this lesson. It is important to note that even though a mathematical topic is listed as a prerequisite, it may in fact be a corequisite, meaning that it can be taught along with the lesson if students have not been exposed to it beforehand.

Technology options are also described to help you make the best use of the technology available to you. In many situations, you will have to use technology to present the material. You simply cannot expect students to iterate a given function enough times by hand in order to determine the ultimate behavior. Most graphing calculators now have "list" or "iterate" features that allow students to compute orbits effortlessly. Some have the web diagram or graphical iteration tool that we use so often in this book. We have found that spreadsheets are ideal tools for iteration as well, not only for their ability to compute orbits quickly, but also for their capability to display simultaneously many of the graphical objects such as time series and histograms that are an essential feature of this material. Spreadsheets also allow for easy changes of parameters in an iteration rule. Finally, there are a number of Java applets available at the Dynamical Systems and Technology Web site **http://math.bu.edu/DYSYS** that were developed specifically for this workbook. These applets run on any computing platform and may be downloaded from this Web site free of charge.

Each section of the book is organized as follows. The *Suggested Lesson Plan* section describes the amount of time needed, possible organizational and presentation strategies, and suggested homework assignments. Keep in mind that these are only suggestions; your own teaching situation may dictate a different structure of time allocation. The *Lesson Notes* contain an assortment of suggestions, ideas, and possible extensions connected with the material in the section.

The **Explanation** pages introduce the mathematical content of the lesson and are intended for you or your students to read. You may want to photocopy the pages for your students, or you may prefer to summarize the concepts using transparencies. In some cases, the material is ideal for student experimentation and individual discovery.

The **Investigations** that follow are in blackline-master format, allowing students to record their answers directly on a photocopy of each page. These are ideal for students to work through cooperatively.

The **Further Exploration** problems extend the ideas from the section. You can photocopy these pages and hand them to students, but they will need to show their work on separate sheets of paper.

Finally, the back of the book contains a thorough and detailed answer to every problem.

Acknowledgments

As with any project that evolves over time, we have many people to thank for their encouragement and assistance. This book grew out of a five-year project sponsored by the National Science Foundation. We are particularly indebted to James Sandefur and Spud Bradley for their support during this period. Many teachers participated in the four-year series of summer workshops that helped to define these materials. We are most appreciative of the work of the lead teachers in these efforts: Beverly Mawn, Jamil Siddiqui, Rob Quaden, Liz Perry, John Bookston, Al Coons, Kathy Leggat, Megan Staples, Jim Carpenter, and Gerald Nimetz. Eileen Lee contributed enormously to the project in the early stages by refining and elaborating on our original notes. Masha Albrecht also contributed a number of fine ideas regarding the organization of these notes. Numerous people contributed to the technology portion of this project, including Clara Bodelon, Rodin Enchev, Noah Goodmann, Kevin Lee, Alex Kasman, Adrian Vajiac, and Johanna Voolich. It is also a pleasure to thank the folks at Key Curriculum Press, especially Steve Rasmussen and David Rasmussen, Jason Luz, James Browne, and Casey FitzSimons for their enthusiasm and helpfulness in bringing this project to a successful conclusion. Finally, we are deeply indebted to Alice Foster, who worked with us for many long hours in the initial stages of this project, trying desperately to keep the mathematical ideas in this book at an appropriate level.

We hope that you find these materials helpful and the mathematical content exciting, stimulating, and challenging. We welcome any comments and/or suggestions about this project. We can be reached via the Web site **http://math.bu.edu/DYSYS**.

Robert L. Devaney
Jonathan Choate

Related Web Site

Check out the Dynamical Systems and Technology Web site at **http://math.bu.edu/DYSYS** for free downloadable Java applets relevant to this book.

Introduction: Why Chaos?

This is a book about one of the most interesting scientific discoveries of recent years, the phenomenon called chaos. Much of what happens in nature can be modeled by mathematical equations. For years scientists have been developing mathematical models for everything from the motion of a simple pendulum to the motion of the planets in the solar system. There are equations for the rise and fall of populations and the ups and downs of the economy. Every field of science and engineering features its own typical mathematical model.

Now, in many cases the mathematical model involved is itself very difficult to solve exactly. Most often, scientists must resort to using computers to approximate the solutions of their mathematical models. Unfortunately, despite major improvements in computational speed and accuracy, scientists have often been unable to make predictions based on the output of the computer. For example, consider the weather. It still seems impossible to accurately forecast the weather one week ahead of time, despite the fact that we can know the current weather at virtually every point on the globe at any given moment.

For years scientists thought that if only they could have access to bigger and better machines, or to faster algorithms, or to more initial data, then they would be able to make accurate predictions. However, in the last 25 years scientists and mathematicians have come to realize that that can never be the case. The culprit is the mathematical phenomenon known as chaos. When chaos is part of a mathematical model, faster and more accurate computing will never lead to complete predictability, for chaos means that very small changes in the initial configuration of the system may lead to great discrepancies down the road. The effect is known as the "butterfly effect," and it is one of the major topics we will discuss.

Now it may seem that such contemporary topics in mathematics would be completely beyond the grasp of high school students. This, however, is not the case. Mathematicians have found that very simple mathematical models can lead to the exact same type of chaos or unpredictability that is observed in weather systems. Indeed, as we shall see below, a mathematical expression as simple as $4x(1 - x)$, when thought of as a dynamical system, can be incredibly chaotic.

To understand this statement, you will have to gather all your mathematical and computational tools. Many of the investigations require technology, usually a graphing calculator or spreadsheet software. Others demand that you be able to interpret very different visual representations of the same mathematical object or set of data. Still others demand that you be able to link your algebraic and geometric skills to tackle different aspects of the same problem. Finally, other investigations require that you use all of these skills at the same time.

So, on the one hand, studying chaos will bring you to the brink of contemporary knowledge about these mathematical systems. On the other hand, it will provide you with a mechanism to use all that you have learned so far in your mathematical career. Good luck!

TEACHER NOTES

Linear Iteration

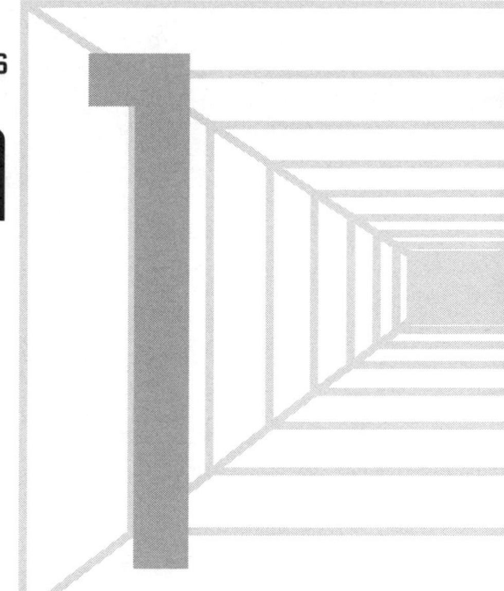

Overview

This lesson is a review of the basic ideas of iteration—iteration rules, orbits, seeds, fixed points, fate of orbits, and time-series graphs. In addition to computing orbits for iteration rules, students determine fixed points by solving simple linear equations. Only linear iteration rules are discussed. Students who have studied *Iteration,* the first book in this series, will find that this section is similar to material presented there and can safely skip ahead. Students who are unfamiliar with iteration will need to spend time here before moving to nonlinear iteration rules.

Mathematical Prerequisites

Students need to be familiar with negative numbers, linear expressions of the form $Ax + B$, solving simple linear equations, and plotting points on a coordinate grid. In Investigation 8, students also need to be familiar with the concept of absolute value.

Mathematical Connections

This lesson has strong connections to the first-year algebra curriculum. The section on **solving for fixed points** is an application of **solving linear equations.** The work with **time series** involves **plotting and interpreting graphs** and illustrates the connections between graphic and symbolic representation of mathematical rules. Students will also work with **substitution** and **simplifying expressions** involving fractions, decimals, and negative numbers.

Further Explorations 5–7 connect directly to the financial problems on savings and loans that are usually included in the second-year algebra curriculum. Further applications on savings, loans, and drug dosing may be found in *Iteration.*

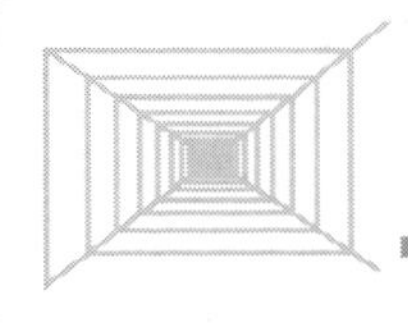

TECHNOLOGY

Although technology is not a prerequisite for this lesson, the iterative feature on many graphing calculators would be helpful to students as they investigate various orbits in the Explanations and Investigations. Of course, several examples should be worked out by hand at first. For the work with time series, either a graphing calculator or spreadsheet software would facilitate demonstrating these graphs and could serve as a tool for students to check graphs drawn manually.

SUGGESTED LESSON PLAN

CLASS TIME

This lesson will require one or two 50-minute class periods depending on how comfortable students are with solving simple linear equations and plotting points. You also might want to consider spending one day introducing the lesson and having students work on the Further Exploration problems in cooperative groups in a second class period.

PREPARATION

Read through the Explanation. Students need to master some terms here: iteration rule, orbit, seed, fixed point, cycle, and time-series graph. Also, arrange for availability of graphing calculators or spreadsheets if you intend to use these either for demonstration or for students to work on problems individually.

LESSON DEVELOPMENT

There are basically three concepts students need to work with in this lesson. The first is the concept of tracing orbits for linear iteration rules given various seeds. The second is plotting time-series graphs in order to visualize orbits geometrically or graphically; the material on time series is important for later use as this is one of the ways that students will view chaos. The third concept involves identifying fixed points by solving a simple linear equation called the fixed-point equation. Advanced classes, or classes on block schedule, should be able to cover all three concepts in one class period. For other classes, you might want to introduce only the first two concepts in one class period, then use a second class period to develop "solving" for fixed points and to look at the special cases for the fixed-point equations.

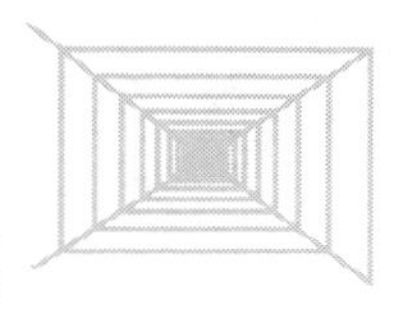

It is important that students be able to articulate the fate of the orbit. There are many different ways that students can explain this. For example, the orbit "gets bigger and bigger," the orbit "tends to infinity," the orbit "just keeps increasing," and so forth, are all acceptable ways of saying the same thing. Have students write a sentence or two about the fate of the orbit in each of the Investigations assigned.

DAY 1

Work out several examples of linear iteration rules by hand. After students have computed several orbits, introduce time-series graphs by suggesting that it might be helpful to visualize the behavior of these orbits by creating a graph. Make sure students understand that these are *not* Cartesian coordinate graphs. Explain that the horizontal axis represents a counter indicating the number of the iteration in the orbit and that the vertical axis is the actual value obtained for that iteration. The graph is keeping track of the series of output values over time—hence the name *time series.*

If time permits, discuss "other fates for orbits" to make students aware that orbits can hop back and forth between positive and negative values. Investigations 3 and 4 involve this type of behavior, so it would be helpful if students have seen an example of this prior to doing the homework.

Assign Investigations 1 and 2 for homework. You could also assign Investigations 3 and 4 and have them skip part a (finding the fixed point) until Day 2.

DAY 2

Have students discuss the results they obtained for the homework Investigations. You might ask them if fixed points can be identified simply by looking at the time-series graph. Then ask them if they can identify fixed points directly from the rule. This provides a nice lead-in for introducing the fixed-point equation, $Ax + B = x$. Ask students why a point (or value of x) would have to satisfy this equation in order to be a fixed point. Have students write and solve the fixed-point equations for Investigations 3 and 4, then check to be sure each solution is in fact a fixed point for that iteration rule. This ties in nicely to the concept of having students always check their solutions to equations to be sure they are correct.

You might also ask students if they can see any reason why it might be helpful to solve for the fixed point before trying to analyze the fates of orbits for an iteration rule. Then have them work on Investigation 5 in class, and they should see that knowing where the fixed point is helps them decide what other

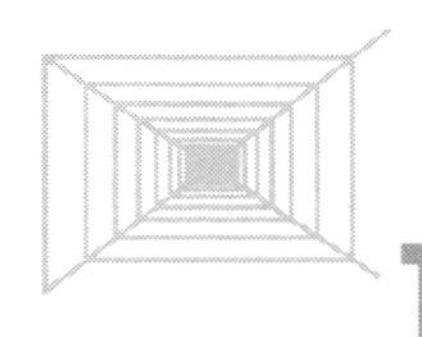

seeds they should try in order to see the possible fates of orbits. This would also be a good time to discuss the special cases of the fixed-point equation and to work on Further Exploration 3.

Assign selected problems from Investigations 5–8 and the Further Exploration problems. Remember that Investigation 8 requires a good understanding of absolute value.

Lesson Notes

Investigations 1–5 all test the same skill: computing orbits and displaying their time-series graphs. However, they all feature different fates of orbits, including going to fixed points (either directly or hopping back and forth), tending to infinity (either directly or hopping back and forth), and tending to infinity without having a fixed point. It is important for students to realize they need to compute orbits on both sides of a fixed point to determine the fates of all orbits. This concept of investigating orbits in various regions is important in preparing students for later investigations involving nonlinear iteration.

Investigation 6 (and Further Exploration 4) is a nice way to have students make the link between time-series graphs and orbits. This will be a recurrent theme in this book: We will derive several different ways of visualizing orbits, and students will need to pass effortlessly between them. This is perhaps the biggest difficulty students will have to overcome in this study, but it is also one of the most worthwhile mathematical skills that they will take away.

Investigations 7–8 are optional. For more information about applications of linear iteration rules, consult *Iteration.*

One of the most important operations in contemporary mathematics and science is **iteration.** To *iterate* means to repeat a process over and over. As we will see in this book, iteration can lead to very complicated mathematical ideas, yet this process has applications in all areas of science, engineering, and social science. For example, ecologists use iteration to study the growth and decline of populations. Economists use iteration to predict how the economy will change, and many of the algorithms that computer scientists use to solve equations involve iteration.

In order to iterate, you need to specify the process that will be repeated (the **iteration rule**) as well as a place to begin the iteration (the **seed**). Sometimes the iteration rule is geometric, and other times it is numerical (that is, given by an algebraic formula). In either case, once you have the rule and the seed you may begin to compute the **orbit** of the iteration rule. The orbit is the list of numbers or geometric figures obtained by successively applying the iteration rule to the output of the previous iteration.

In this lesson you will study the most elementary of all iterations, linear iterations. You will see that the behavior of linear iterations is simple. Don't let this simplicity fool you, however. As you will see very soon, the behavior of nonlinear iteration is quite different.

A NUMERICAL ITERATION RULE

Consider the iteration rule $x \rightarrow x/2 - 2$. Start with the seed 0 for x and apply the rule. The result is -2. Now take this output and apply the rule again. The result this time is -3. Continue this process to generate the orbit

$$0 \rightarrow -2 \rightarrow -3 \rightarrow -3.5 \rightarrow -3.75 \rightarrow -3.875 \rightarrow -3.9375 \rightarrow \cdots$$

Each term in the orbit has a symbol that specifies its position in the orbit. For example, the seed is denoted x_0. So, in this case, $x_0 = 0$. The next term in the orbit is called x_1, so $x_1 = -2$. After five applications of the iteration rule, the term is called x_5. So $x_5 = -3.875$ and $x_6 = -3.9375$.

As you can see, the numbers on this orbit are getting closer and closer to -4.

Continuing, we find

$$x_7 = -3.96875$$

$$x_8 = -3.984375$$

$$x_9 = -3.9921875$$

So we say that the fate of the orbit is: It approaches −4.

If we use the seed −4 for this iteration rule, we find a different fate: The orbit is

$$x_0 = -4$$
$$x_1 = -\frac{4}{2} - 2 = -4$$
$$x_2 = -4$$
$$x_3 = -4$$

and so forth. The orbit stays constant at −4. We therefore say that −4 is a **fixed point** for this iteration rule.

Now start with the seed $x_0 = 2$ and apply the same iteration rule, $x \to x/2 - 2$.

$$2 \to -1 \to -2.5 \to -3.25 \to -3.625 \to -3.8125 \to -3.90625 \to \cdots$$

In this example the orbit seems to get closer and closer to a particular value. What value? The orbit seems to be tending to −4. This was true of the orbit of the seed $x_0 = 0$ as well. Note that this is the fixed point of our iteration rule. It is not a coincidence that both of these orbits tend to the fixed point.

TIME-SERIES GRAPHS

It is helpful to be able to visualize orbits geometrically, not just as a list of numbers. The easiest method to see orbits is to use a **time-series graph.** A time-series graph is a plot of the numbers in the orbit versus the iteration count. That is, on the horizontal axis, we indicate the iteration count 0, 1, 2, ... Along the vertical axis, we place the actual numbers that form the orbit. For example, here is the time-series graph of the iteration rule $x \to x/2 - 2$ with $x_0 = 0$:

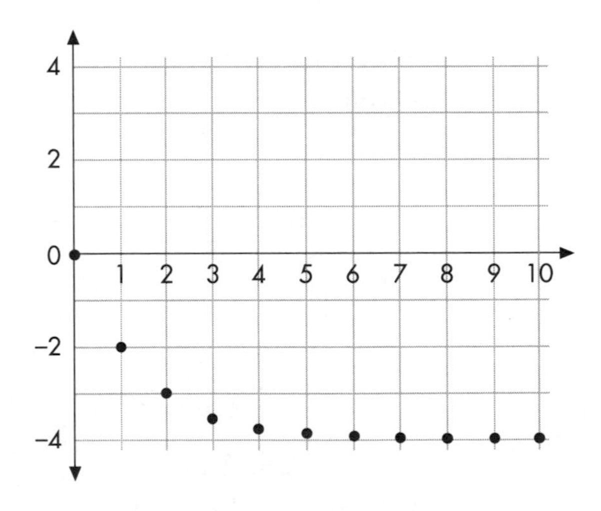

Notice how it is easy to see that the orbit tends to −4 using the time-series graph.

LINEAR ITERATION RULES

An iteration rule of the form $x \to Ax + B$ where A and B are constants is called a **linear iteration rule.** Linear iteration rules occur often in applications. For example, bank savings plans that compound interest each year assume the form $x \to (1 + r)x + P$ where both r (the annual interest rate) and P (the annual deposit) are constants. (See the Further Exploration problems.)

Our job in this section is to understand the fates of any orbit under a linear iteration rule. We already looked at a linear iteration rule $x \to x/2 - 2$ that has orbits that tend to a fixed point. Now consider the iteration rule $x \to 2x + 1$. The orbit of the seed $x_0 = 0$ is

$$0 \to 1 \to 3 \to 7 \to 15 \to 31 \to 63 \to \cdots$$

Clearly, the numbers on this orbit grow larger and larger. We say that this orbit **tends to infinity**. Similarly, the orbit of $x_0 = -2$ tends to (negative) infinity since the orbit is

$$-2 \to -3 \to -5 \to -9 \to -17 \to -33 \to \cdots$$

Finally, the orbit of $x_0 = -1$ is fixed since $x_1 = 2(-1) + 1 = -1$.

You can view these three orbits in a time-series graph that looks like this:

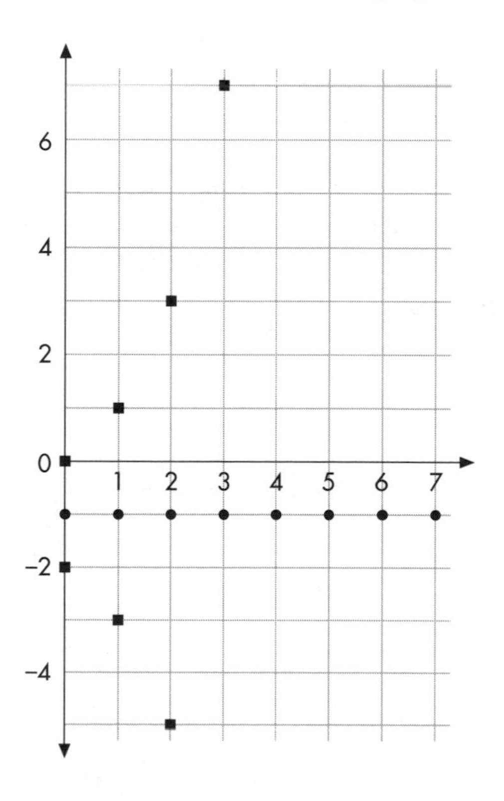

OTHER FATES FOR ORBITS

As the preceding examples show, a linear iteration rule may have orbits that are fixed, that tend to fixed points, or that tend to infinity. There are several other types of orbits that occur. First, consider the iteration rule $x \to -2x$. Clearly, 0 is a fixed point for this iteration rule, but try any other seed. All other orbits eventually alternate between large positive and large negative values. For example, the orbit of $x_0 = 2$ under this rule begins

$$2 \to -4 \to 8 \to -16 \to 32 \to -64 \to 128 \to -256 \to \cdots$$

So we say that this orbit tends to positive and negative infinity.

Next consider the iteration rule $x \to -x/2$. For this rule, 0 is again a fixed point. This time, all other orbits tend to 0, but they do so by hopping back and forth on either side of 0. For example, the orbit of $x_0 = 2$ is

$$2 \to -1 \to 0.5 \to -0.25 \to 0.125 \to -0.0625 \to 0.03125 \to \cdots$$

Finally, orbits may cycle. Look at the orbit of $x_0 = 4$ for the linear iteration rule $x \to -x + 2$:

$$4 \to -2 \to 4 \to -2 \to 4 \to -2 \to \cdots$$

This orbit is a **cycle of period 2** since the orbit repeats every second iteration.

FINDING FIXED POINTS

Fixed points play a very important role in the study of iteration. Luckily, for linear iteration rules, we can find fixed points in a straightforward way. If x_0 is a fixed point for the iteration rule $x \to Ax + B$, then x_0 must satisfy the **fixed-point equation** $Ax + B = x$. Using a little algebra, we can find the solution of this equation:

$$Ax - x = -B$$

$$(A - 1)x = -B$$

$$x = \frac{-B}{A - 1}$$

Remember that we know A and B, so we know the value of the right side of this equation explicitly.

This right side is fine as long as $A \neq 1$. In that case, we have 0 in the denominator, so we have to watch out. As long as $A \neq 1$, we now know that our linear iteration rule $x \rightarrow Ax + B$ has a fixed point located at $x = {}^{-B}/_{(A-1)}$ and, moreover, that this is the only fixed point. For example, the iteration rule

$$x \rightarrow \frac{1}{3}x + 2$$

has a fixed point at $x = 3$, since $A = 1/3$ and $B = 2$, so

$$x = \frac{-2}{\frac{1}{3} - 1} = \frac{-2}{-\frac{2}{3}} = -2 \cdot \left(-\frac{3}{2}\right) = 3.$$

As a check, for the seed $x_0 = 3$, we have $x_1 = 1/3(3) + 2 = 3 = x_0$, so 3 is indeed a fixed point.

When $A = 1$, we have a special case. We have the iteration rule $x \rightarrow x + B$. If we try to solve the fixed-point equation $x + B = x$, we find that after subtracting x from both sides of the equation, we must have $B = 0$. So that means the iteration rule $x \rightarrow x + B$ has no fixed points when $B \neq 0$. Of course, that is pretty clear, since the orbit of x is given by

$$x \rightarrow x + B \rightarrow x + 2B \rightarrow x + 3B \rightarrow \cdots$$

and so on. This orbit either tends to infinity (if $B > 0$) or tends to negative infinity (if $B < 0$). Only when $B = 0$ do we find a fixed point, and in that case, every x is a fixed point. That's because our iteration rule is $x \rightarrow x$, which is not an exciting iteration rule. Nothing moves at all!

SUMMARY

Let's summarize all of this. The linear iteration rule $x \rightarrow Ax + B$ always has a single fixed point at

$$x = \frac{-B}{A - 1}$$

except in two special cases: When $A = 1$ and $B \neq 0$, there are no fixed points, and when $A = 1$ and $B = 0$, every x is a fixed point.

NAME(S):

1 ▹ COMPUTING ORBITS

Consider the linear iteration rule $x \rightarrow x/2 + 2$.

a. Describe the fate of the orbit for each seed in the table below.

Seed	**Fate of orbit**
$x_0 = 0$	Orbit tends to 4
$x_0 = 6$	
$x_0 = 8$	
$x_0 = -8$	
$x_0 = -10$	
$x_0 = -4$	
$x_0 = 4$	

b. What appears to be true about the fate of any orbit for this rule?

____

____

c. Try a few more seeds to verify your conjecture from part b.

Seed	**Fate of orbit**
$x_0 =$	
$x_0 =$	
$x_0 =$	

d. Was your conjecture confirmed?

e. Sketch the time-series graph for the orbits of the seeds $x_0 = 0$, $x_0 = 4$, $x_0 = 8$, and $x_0 = -4$.

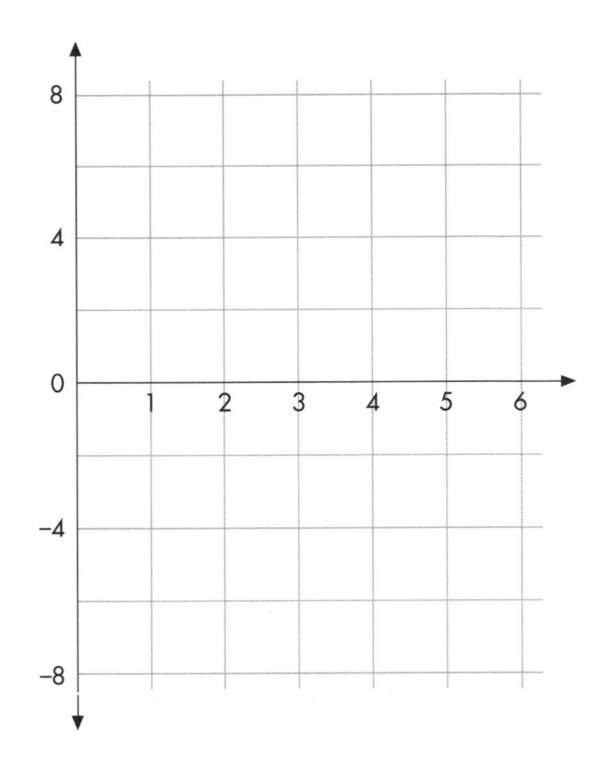

2 ▹ ANOTHER ORBIT

Consider the linear iteration rule $x \rightarrow 2x - 2$.

a. Describe the fate of the orbit for each seed in the table at the right.

Seed	Fate of orbit
$x_0 = 0$	Orbit tends to (negative) infinity
$x_0 = 1$	
$x_0 = 3$	
$x_0 = -5$	
$x_0 = 5$	
$x_0 = -2$	
$x_0 = 2$	

b. What do you suspect is happening for this iteration rule?

____

____

c. Do all orbits behave similarly?

____

____

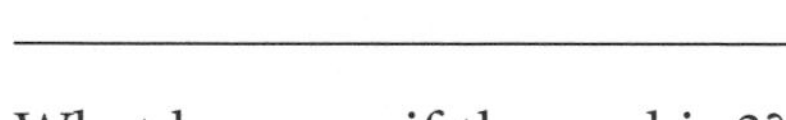

d. What happens if the seed is 2?

____

____

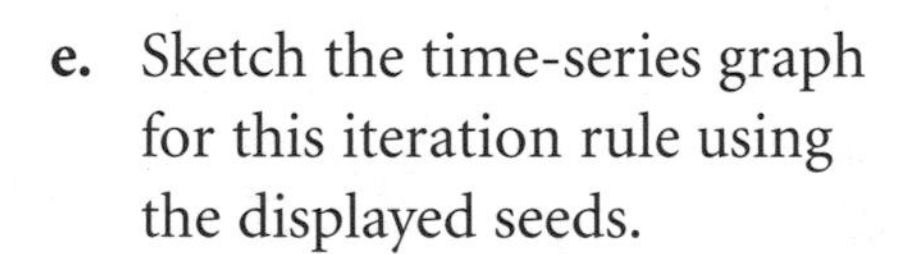

e. Sketch the time-series graph for this iteration rule using the displayed seeds.

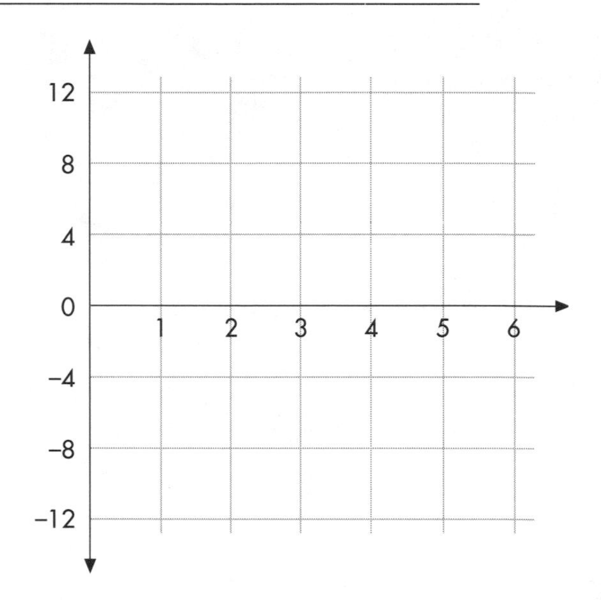

3 ▹ One more orbit

Consider the linear iteration rule $x \rightarrow -2x + 3$.

a. Find the fixed point for this iteration.

b. Describe the fate of the orbit for each seed in the table at the right.

c. What appears to be true about the fate of any orbit for this rule?

____

____

____

____

Seed	Fate of orbit
$x_0 = 0$	
$x_0 = -1$	
$x_0 = 10$	
$x_0 = -6$	
$x_0 = -10$	
$x_0 = -\frac{1}{4}$	
$x_0 = \frac{1}{8}$	

4 ▹ STILL ANOTHER ORBIT

Consider the linear iteration rule $x \to -x/2 + 3$.

a. Find the fixed point for this iteration.

b. Describe the fate of the orbit for each seed in the table below.

Seed	Fate of orbit
$x_0 = 0$	
$x_0 = -4$	
$x_0 = 10$	
$x_0 = 4$	
$x_0 = -10$	
$x_0 = 100$	
$x_0 = -46.3$	

c. What appears to be true about the fate of any orbit for this rule?

____

____

5 ▹ YOU CHOOSE THE SEEDS

Consider the iteration rule $x \rightarrow -x + 1$.

a. Find the fixed point for this iteration rule.

b. Choose four seeds, compute their orbits, and display the time-series graphs here:

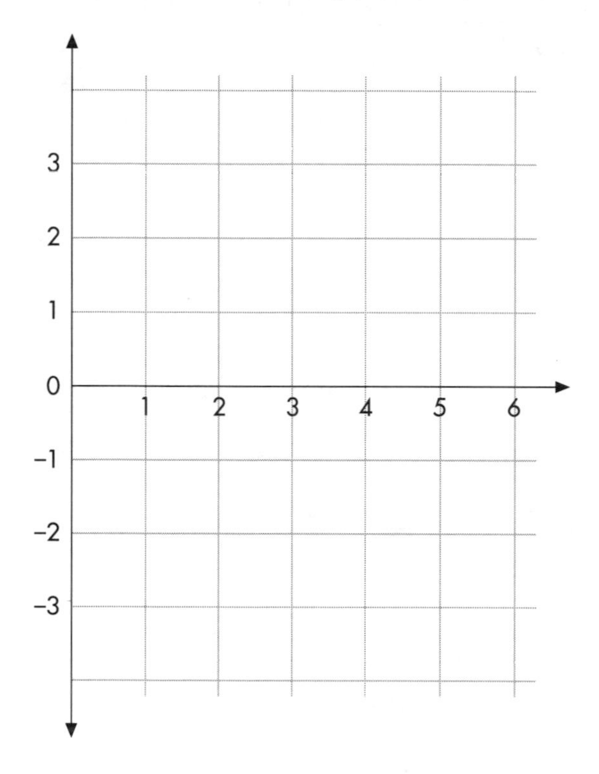

c. What appears to be true about the fate of any orbit for this rule?

6 ▹ Matching orbits and time-series graphs

Here are five portions of orbits and five time-series graphs. Match the orbits with the time-series graphs. Note that we do not put any numbers on the axes; you have to use the relative positions of the numbers on the orbit to find the correct time-series graph.

a. 0, 2, 6, 14 matches graph ____________.

b. 0, –6, –8, $-\frac{28}{3}$ matches graph ____________.

c. 0, 1, 2, 3 matches graph ____________.

d. 0, –6, –4, $-\frac{14}{3}$ matches graph ____________.

e. 0, –4, 0, –4 matches graph ____________.

A.

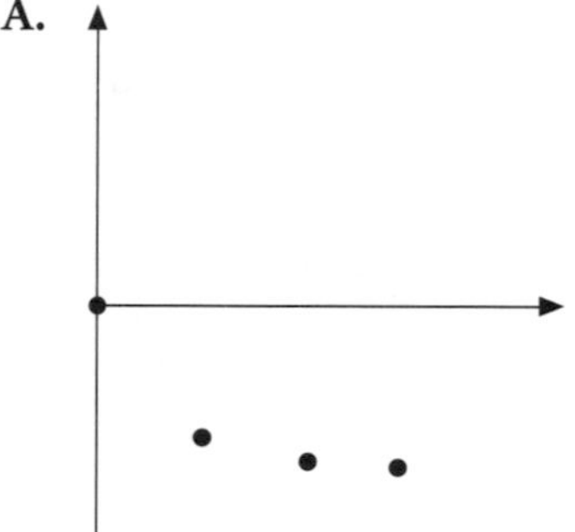

B.

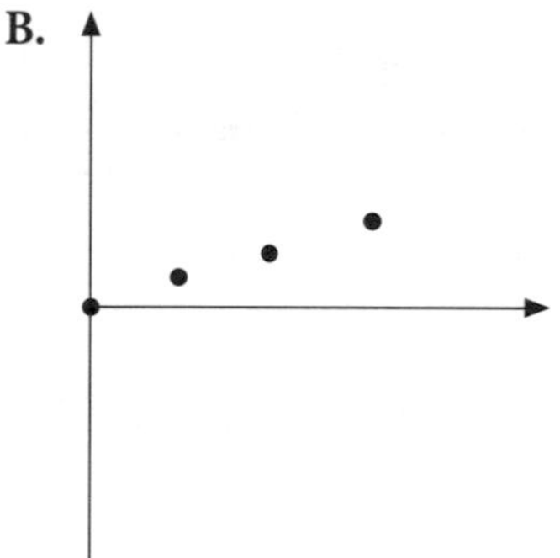

C.

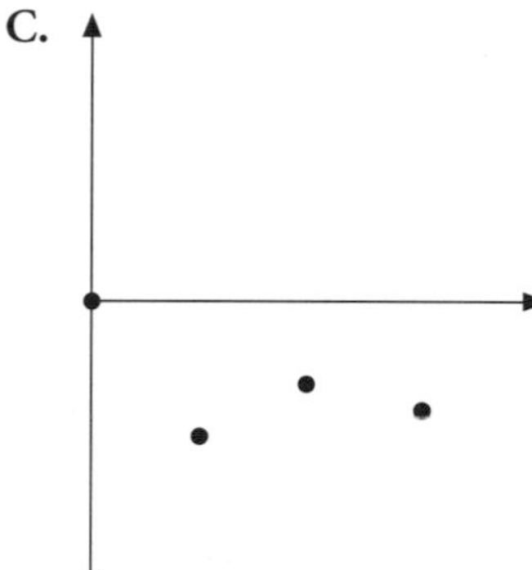

D.

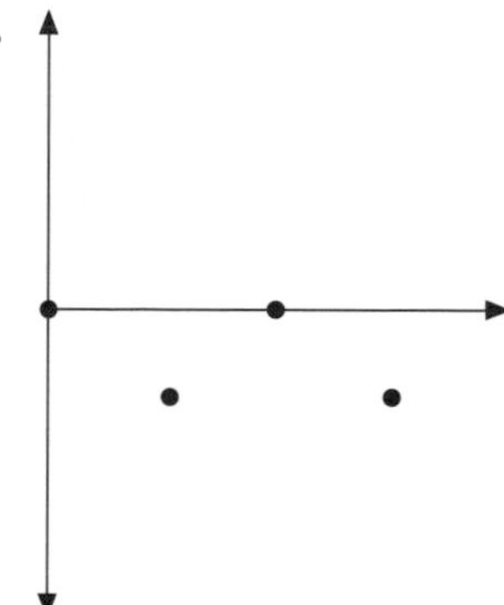

E.

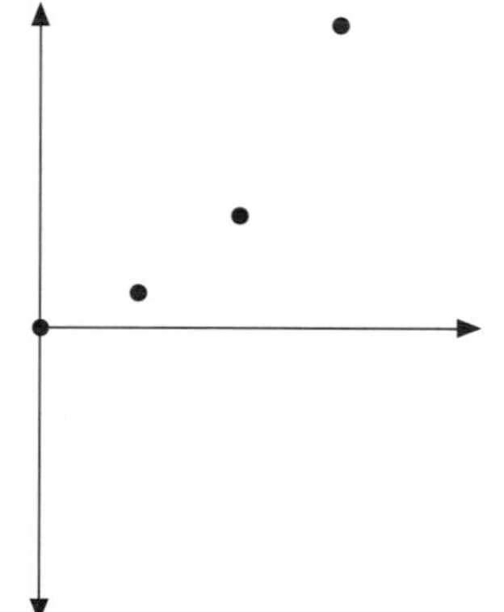

7 ▹ Be careful with this one!

Here are seven portions of orbits. Which of them could be orbits arising from a linear iteration rule? Explain your reasons.

a. 0, 0, 0, 1, 2, 3, . . .

b. 9, 6, 3, 0, –3, . . .

c. 0, 1, 2, –1, –2, –3, . . .

d. 73, 73, 73, 73, . . .

e. 4, 7, 13, 25, 49, . . .

f. 0, –3, –9, –21, –66, . . .

g. 5, 0, 5, 0, 5, 0, . . .

8 ▹ Piecewise linear iteration rules

An iteration rule of the form $x \to A|x| + B$ is called a **piecewise linear iteration rule** because the rule comes in two linear pieces: one for $x > 0$ and the other for $x < 0$. Describe the fate of all orbits for each iteration.

a. $x \to |x|$

b. $x \to |x| + 1$

c. $x \to 2|x|$

1. Discuss the fate of all orbits for these linear iteration rules. First find the fixed point for the rule. Then calculate at least six different orbits to support your conclusion.

   a. $x \rightarrow 3x - 2$

   b. $x \rightarrow -\frac{x}{3} - 4$

2. Can a linear iteration rule ever have more than one fixed point? Explain your answer.

____

____

3. Here are seven time-series graphs, each of which displays two orbits from a linear iteration rule $x \rightarrow Ax + B$. Identify which time-series graph corresponds to a rule with these values of A:

   a. $A = -1$

   b. $0 < A < 1$

   c. $A < -1$

   d. $-1 < A < 0$

   e. $A > 1$

   f. $A = 1$

   g. $A = 0$

A.

B.

C.

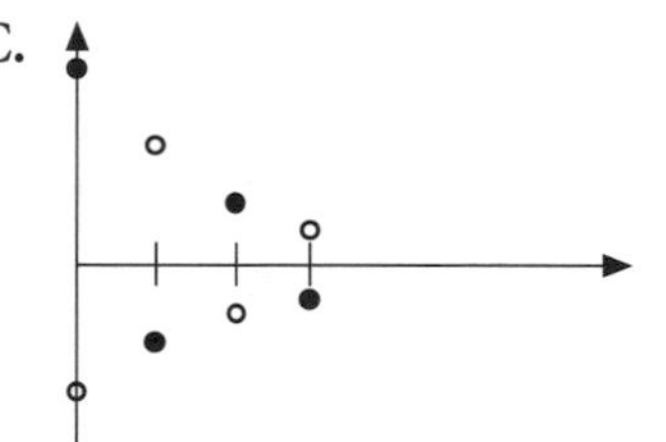

D.

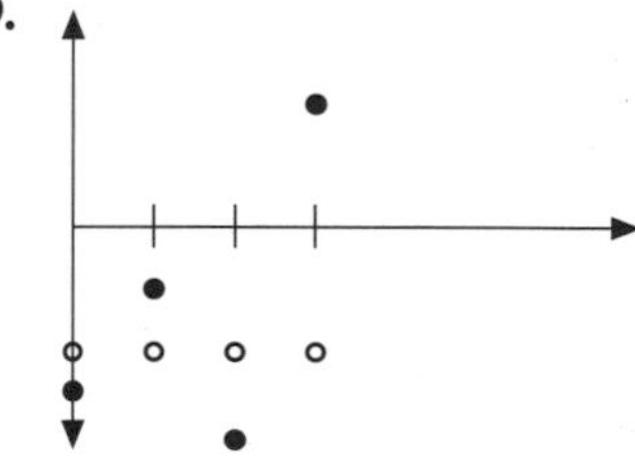

E.

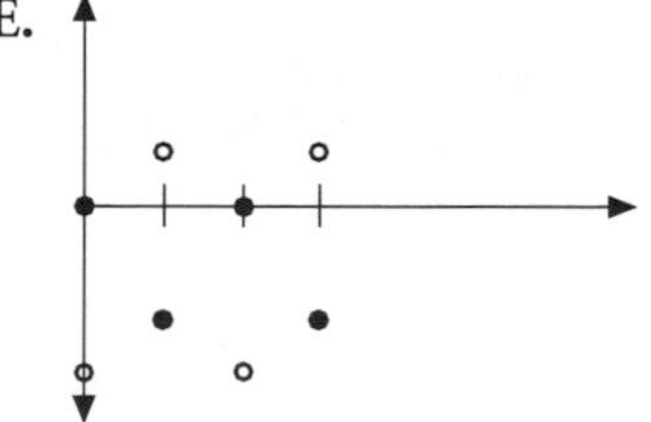

F.

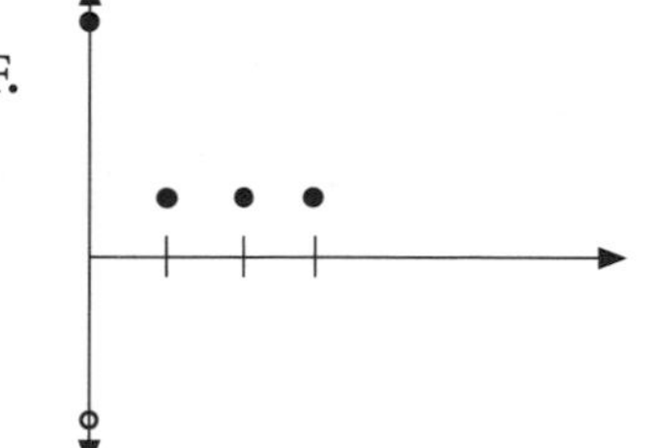

G.

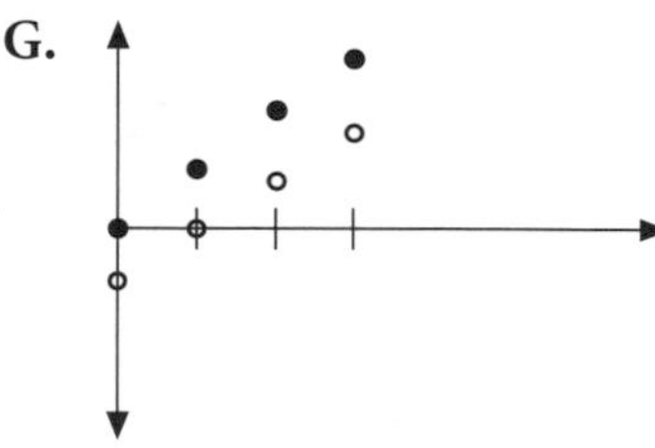

4. Identify the linear iteration rules (that is, determine A and B) that generate each of the following orbits. *Caution:* There may be more than one possible rule!

a. $0, -3, 0, -3, \ldots$

b. $3, 3, 3, 3, \ldots$

c. $2, 4, 6, 8, 10, \ldots$

d. $16, -8, 4, -2, 1, \ldots$

e. $5, 3, 2, 1.5, 1.25, \ldots$

f. $-3, 2, 2, 2, 2, \ldots$

5. Suppose you make an initial deposit of \$100 in a savings account with an annual interest rate of 5%. Write a linear iteration rule that gives the account balance after each succeeding year. How much will you have on deposit after 5 years? After 10 years?

6. Suppose you make an initial deposit of \$100 and an additional deposit of \$100 every year thereafter in a savings account with an annual interest rate of 8%. Write a linear iteration rule that gives the account balance after each succeeding year. How much will you have on deposit after 5 years? After 10 years?

7. Suppose that the interest rate in a savings account is 8% annually and that your initial deposit is \$2000. If you add \$500 to this account at the end of each year, what is your balance at the end of the first 5 years? At this rate of saving, how long will it take for you to have enough money in the bank to purchase a \$9000 car? Do you think that you should look for a cheaper car?

TEACHER NOTES

Types of Fixed Points

Overview

The main goal of this lesson is to have students understand the difference between various types of fixed points: attracting, repelling, and neutral. They need to develop a good comprehension of these terms in preparation for later lessons. In addition to investigating the three types of fixed points through numerical iteration, students plot time-series graphs for each of the major types of fixed points.

Mathematical Prerequisites

Students need to understand iteration and how to compute orbits in order to understand the difference between attracting, repelling, and neutral fixed points.

Mathematical Connections

Topics that relate to the concepts in this lesson include **solving linear equations, plotting time-series graphs,** and **sequences.** The idea of attracting fixed points also connects to the concept of **asymptotic behavior.** This might also be a good place to refer to the distinction between **continuous** and **discrete graphs** as students plot time-series graphs.

Technology

As in the previous lesson, although technology is not a necessity, calculators or spreadsheets are useful tools for performing any iteration in this lesson and for helping students determine whether a fixed point is attracting, repelling, or neutral. Again, a few fixed points should be investigated by hand before turning to technology.

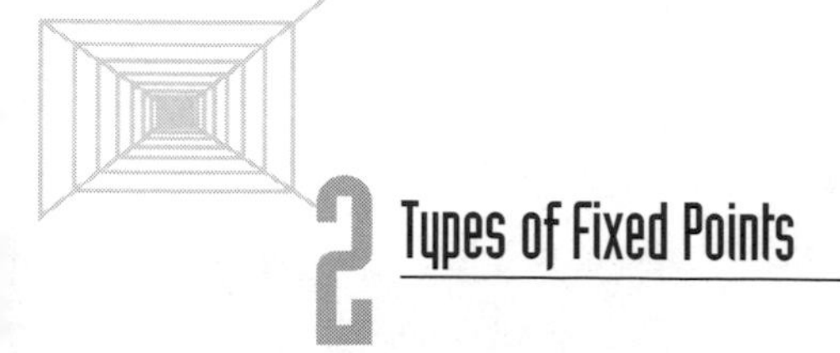

SUGGESTED LESSON PLAN

CLASS TIME

One 50-minute class period should be devoted to developing the classification of fixed points of linear iteration rules and looking at the associated time-series graphs. The main goal is to get students to understand that a fixed point of a linear iteration rule $x \rightarrow Ax + B$ is attracting if $|A| < 1$, repelling if $|A| > 1$, and neutral if $|A| = 1$. Perhaps a summary of the rule can be made during the next class period after students have worked some of the Investigations.

PREPARATION

The main goal is to have students discover the classification of fixed points. This is contained in Investigation 7. Even if students do not come up with this classification, it is important to cover this in class since it relates to a similar classification for nonlinear iteration rules.

LESSON DEVELOPMENT

Cover the introductory material dealing with attracting, repelling, and neutral fixed points. Have students investigate several rules and a variety of seeds (as in Investigations 1–5) so they develop a solid understanding of the differences between the types of fixed points. Students should look at the time-series graphs for each of these cases in addition to computing the numerical orbits. They should work Investigations 1–5 and possibly Investigation 6 before trying to come up with the main classification in Investigation 7.

Investigation 9 is a very good (and very challenging!) way for students to begin to visualize the relationship between slope and the time-series graph.

LESSON NOTES

Make sure students realize that the initial examples for introducing attracting fixed points, repelling fixed points, and neutral points do not prove that the points are of a particular type; rather, they are enough evidence to lead to a conjecture. In the next lesson we will give a more convincing graphical technique for seeing why these fixed points behave as they do.

Investigations 8–10 are more challenging. For students who plan to move on to *The Mandelbrot and Julia Sets,* the fourth book in this series, Investigation 10 is a good lead-in to the notion of a parameter plane, developed more fully in Investigation 11.

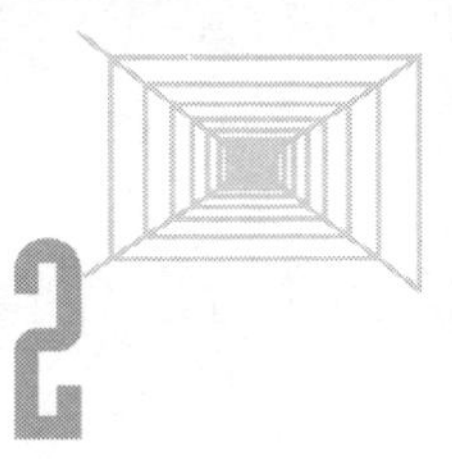

Note that we have not discussed the case $A = 0$ in this lesson. For this iteration rule $x \rightarrow B$, the point $x = B$ is a fixed point, and all other orbits are eventually fixed at B (they are sent to B after one iteration). Some people say that B is an attracting fixed point, but others prefer to call this rule a special case. There is no real agreement, both are correct, and you can handle this issue as you see fit. This is a good topic for class discussion.

As we have seen, linear iteration rules of the form $x \to Ax + B$ always have a fixed point (unless $A = 1$). In this lesson we will classify fixed points according to the behavior of other orbits.

ATTRACTING FIXED POINTS

As a first example, consider the linear iteration rule $x \to x/2 - 1$. For this rule, $x_0 = -2$ is a fixed point. Try it! Use -2 as the seed, and put it into the iteration rule.

If we consider the orbit of the seed $x_0 = 2$, we find that the orbit is

$$2 \to 0 \to -1 \to -1.5 \to -1.75 \to -1.875 \to \cdots \to -1.9990\ldots \to -1.9995\ldots$$

This orbit seems to be tending toward the fixed point at $x_0 = -2$. If we consider the orbit of $x_0 = 17$, a similar result occurs:

$$17 \to 7.5 \to 2.75 \to 0.375 \to -0.8125 \to 1.40625 \to \cdots \to -1.9814\ldots \to \cdots$$

In the table on the following page, look at five more orbits for this iteration rule. Here we list the seeds in the top row. The orbit for each seed is directly below it. The iteration count appears in the left column.

We say that a fixed point for the linear iteration rule $x \to Ax + B$ is **attracting** if all other orbits tend to the fixed point under iteration. Of course, the table does not guarantee that *all* orbits tend to the fixed point at -2; it is only numerical evidence for this fact. To verify that -2 is an attracting fixed point, we would need to do some additional algebra.

Count	5 different seeds				
0	100	25	−50	−10	6
1	49	11.5	−26	−6	2
2	23.5	4.75	−14	−4	0
3	10.75	1.375	−8	−3	−1
4	4.375	−0.3125	−5	−2.5	−1.5
5	1.1875	−1.1563	−3.5	−2.25	−1.75
6	−0.4063	−1.5781	−2.75	−2.125	−1.875
7	−1.2031	−1.7891	−2.375	−2.0625	−1.9375
8	−1.6016	−1.8945	−2.1875	−2.0313	−1.9688
9	−1.8008	−1.9473	−2.0938	−2.0156	−1.9844
10	−1.9004	−1.9736	−2.0469	−2.0078	−1.9922
11	−1.9502	−1.9868	−2.0234	−2.0039	−1.9961
12	−1.9751	−1.9934	−2.0117	−2.002	−1.998
13	−1.9875	−1.9967	−2.0059	−2.001	−1.999
14	−1.9938	−1.9984	−2.0029	−2.0005	−1.9995
15	−1.9969	−1.9992	−2.0015	−2.0002	−1.9998

REPELLING FIXED POINTS

There are other types of fixed points for linear iterations. A fixed point is **repelling** if all other orbits tend to move away from the fixed point under iteration. For example, consider the rule $x \rightarrow -3x + 2$. This iteration has a fixed point at $x_0 = 0.5$, but the orbits of nearby seeds tend away from this point. For example, we list the first few iterations of 0.6, 0.4, 0.51, and 0.49 in the following table:

Count	5 different seeds			
0	0.6	0.4	0.51	0.49
1	0.2	0.8	0.47	0.53
2	1.4	−0.4	0.59	0.41
3	−2.2	3.2	0.23	0.77
4	8.6	−7.6	1.31	−0.31
5	−23.8	24.8	−1.93	2.93
6	73.4	−72.4	7.79	−6.79
7	−218.2	219.2	−21.37	22.37
8	656.6	−655.6	66.11	−65.11
9	−1967.8	1968.8	−196.33	197.33
10	5905.4	−5904.4	590.99	−589.99
11	−17714.2	17715.2	−1771	1772
12	53144.6	−53143.6	5314.9	−5313.9

Notice that each orbit moves away from the fixed point and then tends to positive and negative infinity. Again this is only numerical evidence, not a proof, but we do see that seeds that are very close to the fixed point have orbits that eventually move very far away.

Neutral fixed points

There is another type of fixed point. For example, the linear iteration rule $x \to -x + 2$ has a fixed point at $x_0 = 1$, but it is neither attracting nor repelling.

For example, the orbit of $x_0 = 7$ is

$$7 \to -5 \to 7 \to -5 \to \cdots$$

and the orbit of $x_0 = -1$ is

$$-1 \to 3 \to -1 \to 3 \to \cdots$$

Indeed, if we choose any other seed x_0, then the orbit is

$$x_0 \to -x_0 + 2 \to -(-x_0 + 2) + 2 = x_0$$

so the orbit returns to itself after two iterations. That is, the orbit starts to cycle after two iterations. So all other orbits lie on a **2-cycle.** We also say that this orbit is a **cycle of period 2.** We say that a fixed point that is neither attracting nor repelling is a **neutral** fixed point. As another example, the simple iteration rule $x \to x$ has the property that any seed is a fixed point. So orbits are neither attracted to nor repelled from any fixed point, and each fixed point is neutral.

NAME(S):

1 ▹ A LINEAR ITERATION RULE

Consider the iteration rule $x \to x/2 + 1$.

a. Find the fixed point for this rule. ______________________

b. Find the orbit for each seed.

$x_0 = 4$

____

____

$x_0 = -4$

____

____

$x_0 = -1$

____

____

$x_0 = 6$

____

____

c. Sketch the time-series graph for all of these orbits on the same graph.

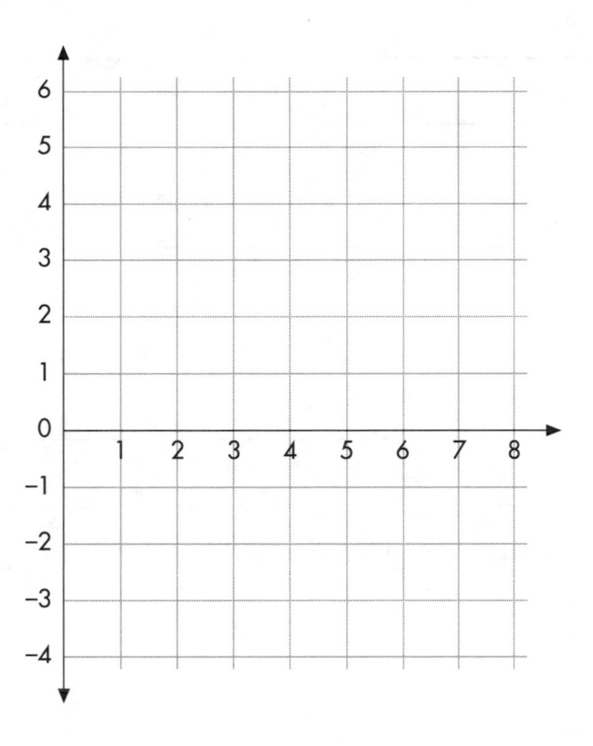

d. Make a conjecture about the fate of all orbits for this iteration rule.

____

____

c. Is the fixed point attracting, repelling, or neutral? ____________

2 ▹ ANOTHER LINEAR ITERATION RULE

Consider the iteration rule $x \rightarrow -2x + 3$.

a. Find the fixed point for this rule. ____________

b. Find the orbit for each seed.

$x_0 = 4$

____

____

$x_0 = -4$

$x_0 = -1$

$x_0 = 0$

c. Sketch the time-series graph for all of these orbits on the same graph.

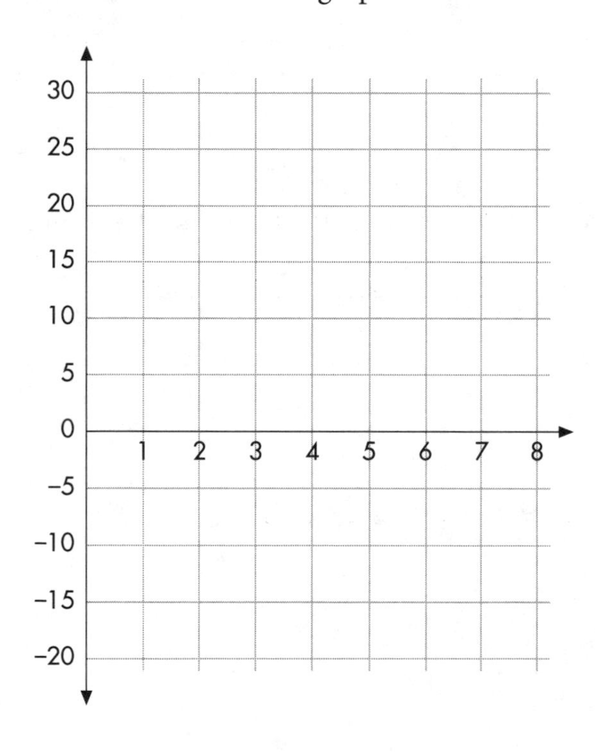

d. Make a conjecture about the fate of all orbits for this iteration rule.

____

____

e. Is the fixed point attracting, repelling, or neutral? ______________

3 ▹ A THIRD LINEAR ITERATION RULE

Consider the iteration rule $x \rightarrow 3x/2 + 1$.

a. Find the fixed point for this rule. ______________

b. Find the orbit for each seed.

$x_0 = 4$

____

____

$x_0 = -4$

____

____

$x_0 = -1$

____

____

$x_0 = 0$

____

____

c. Sketch the time-series graph for all of these orbits on the same graph.

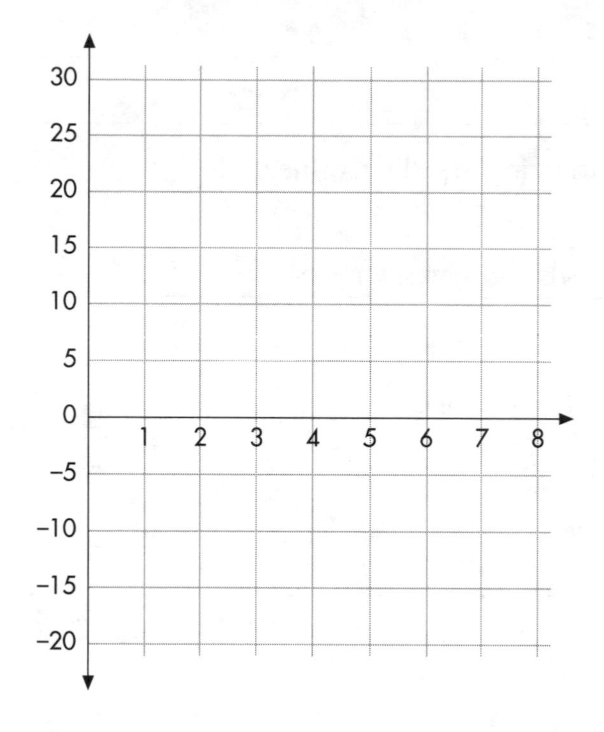

d. Make a conjecture about the fate of all orbits for this iteration rule.

____

____

e. Is the fixed point attracting, repelling, or neutral? ________________

4 ▹ ONE MORE LINEAR ITERATION RULE

Consider the iteration rule $x \rightarrow -x + 6$.

a. Find the fixed point for this rule. ________________

b. Find the orbit for each seed.

$x_0 = 14$

____

____

$x_0 = -20$

____

____

$x_0 = -1$

____

____

$x_0 = 0$

____

____

c. Sketch the time-series graph for all of these orbits on the same graph.

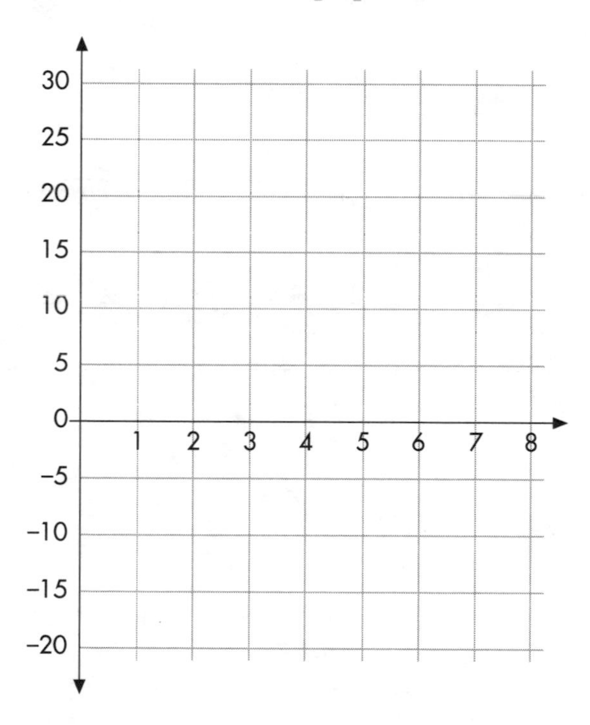

d. Make a conjecture about the fate of all orbits for this iteration rule.

____

____

e. Is the fixed point attracting, repelling, or neutral? ____________________

5 ▹ AND ANOTHER . . .

Consider the iteration rule $x \rightarrow -0.6x$.

a. Find the fixed point for this rule.

b. Find the orbit for each seed.

$x_0 = 40$

$x_0 = -20$

$x_0 = -12$

$x_0 = 6$

c. Sketch the time-series graph for all of these orbits on the same graph.

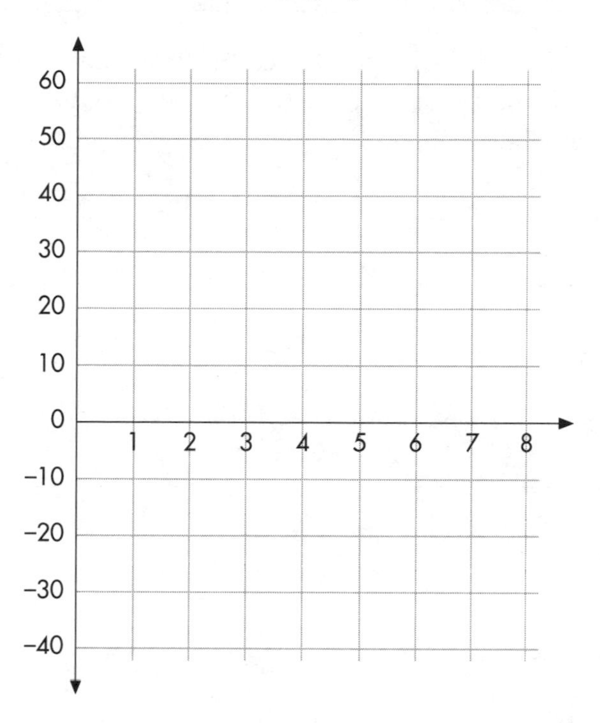

d. Make a conjecture about the fate of all orbits for this iteration rule.

____

____

e. Is the fixed point attracting, repelling, or neutral? ________________

6 ▸ DETERMINING THE TYPES OF FIXED POINTS

For each of the following iteration rules, determine the fixed point and decide if the fixed point is attracting, repelling, or neutral.

a. $x \rightarrow 4x + 2$ ____________________________

b. $x \rightarrow \frac{x}{4} + 2$ ____________________________

c. $x \rightarrow -x + 20$ ____________________________

d. $x \rightarrow 0.2x + 12$ ____________________________

7 ▹ YOUR FIXED-POINT CONJECTURE

Use the results of the previous investigations to conjecture about the types of fixed points for a linear iteration rule of the form $x \rightarrow Ax + B$. When is the fixed point attracting, repelling, or neutral?

____

____

____

____

____

____

8 ▹ A TIME-SERIES GRAPHICAL SUMMARY

Consider the linear iteration rule $x \rightarrow Ax + B$ with fixed point P_0, where we have not specified P_0, A, and B exactly. Draw the approximate locations of the next five points in the time series for each of x_0 and y_0 in the three cases below.

a. $0 < A < 1$

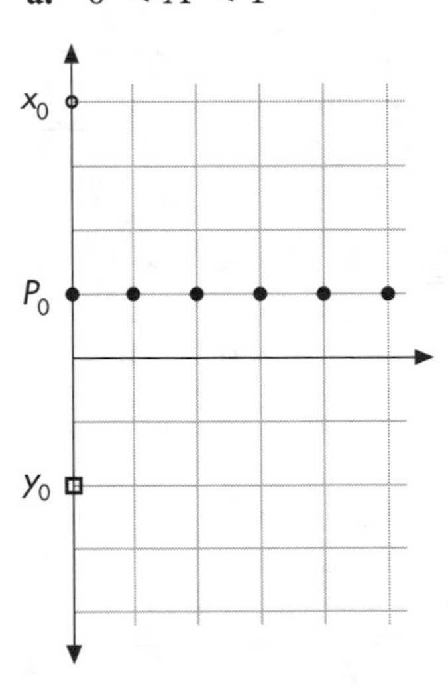

b. $-1 < A < 0$

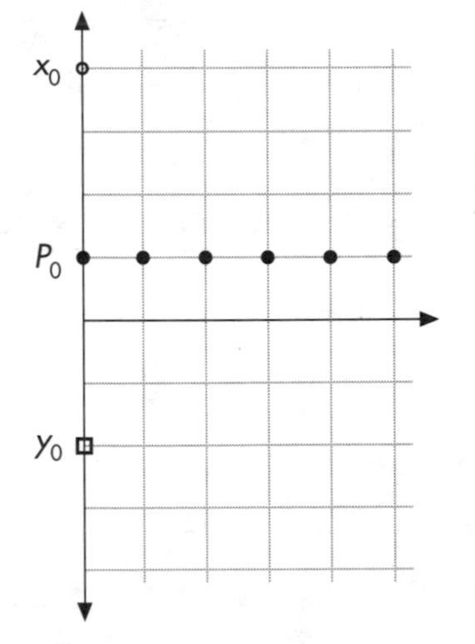

c. $A = -1$

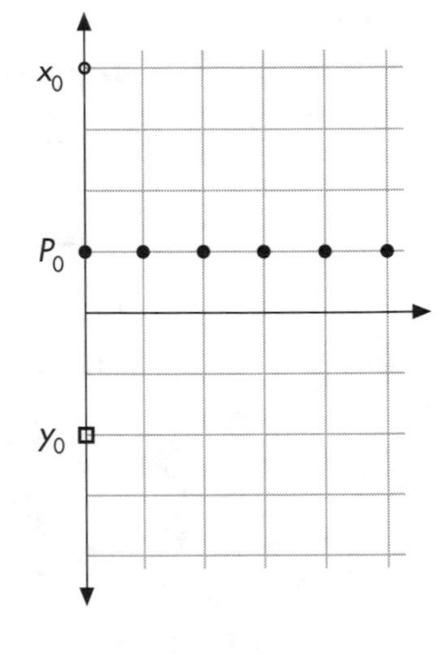

9 ▹ FIXED POINT FREE

For which values of A and B does the linear iteration rule $x \rightarrow Ax + B$ have no fixed points whatsoever? ______________________________

10 ▹ MATCHING TIME-SERIES GRAPHS

Here is a thought-provoking exercise. Each time series corresponds to a different linear iteration rule of the form $x \rightarrow Ax + B$. Decide which time series goes with which A and B value pair given below. Unfortunately, we are not going to tell you the exact values of A and B. Worse, in some cases you will first have to figure out the approximate location of the fixed point before determining the answer.

a. $A = 1, B > 0$ **b.** $A = 1, B < 0$ **c.** $-1 < A < 0, B < 0$

d. $A > 1, B > 0$ **e.** $A = 0, B > 0$ **f.** $A < -1, B > 0$

g. $0 < A < 1, B < 0$ **h.** $A > 1, B < 0$ **i.** $A = -1, B = 0$

A.

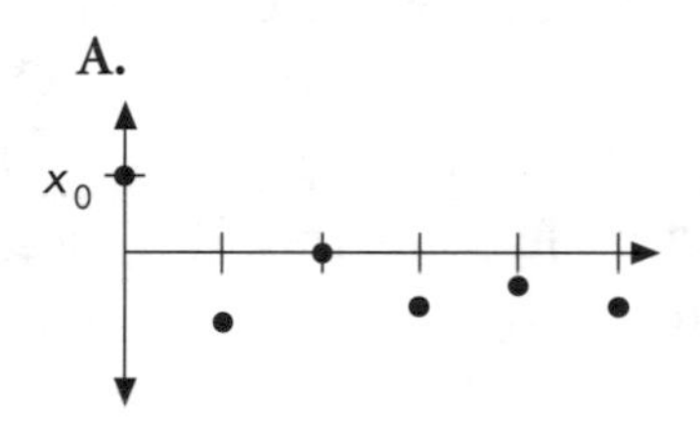

B.

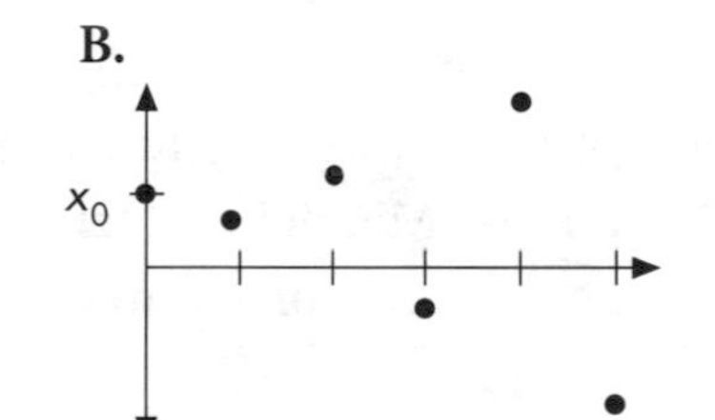

C.

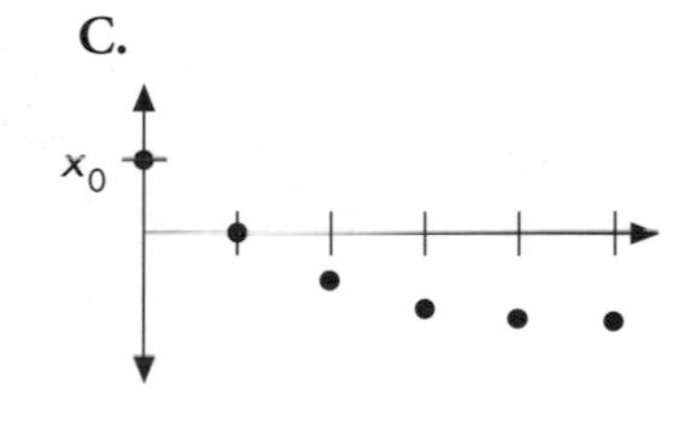

D.

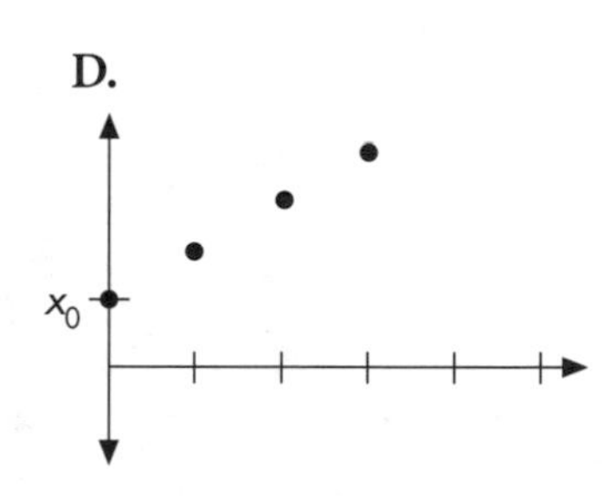

E.

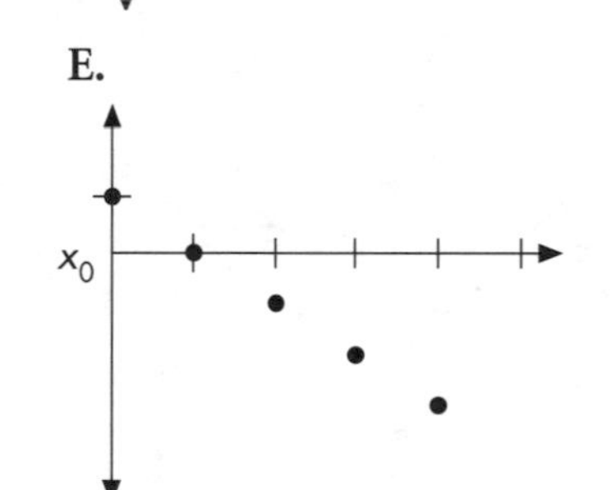

F.

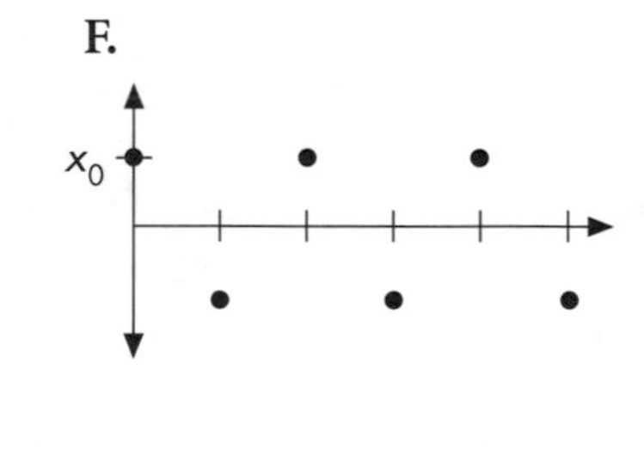

G.

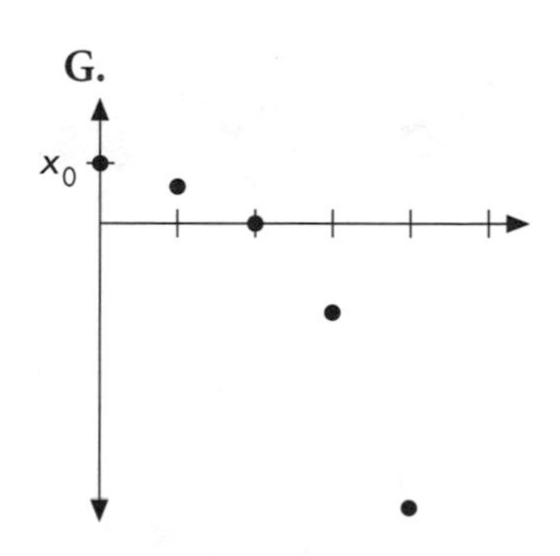

11 ▹ THE BIG PICTURE

You will now paint the picture of the **parameter plane** for linear iterations. This is a record of all the possible behaviors of a linear iteration rule. On a piece of paper, draw the *AB*-plane. Let the *A*-axis be horizontal and the *B*-axis vertical. You will color each point (A, B) in this plane with a color that depends on the qualitative behavior of the linear iteration rule $x \rightarrow Ax + B$. For example, you might color each (A, B) red if the iteration rule has an attracting fixed point for which orbits do not hop back and forth over the fixed point. If the fixed point is attracting but the orbits hop back and forth about the fixed point, color (A, B) a different color. Remember to "paint" all the special cases in a different color. Include a table at the bottom of your picture that records what each different color means. (For this exercise, it is only the qualitative behavior that matters, not the exact location of the fixed point.)

The numbers *A* and *B* are called **parameters.** They are numbers that you control; they determine which linear iteration rule you are using. Once you have selected your parameters, you begin to iterate and discover certain behaviors for the orbits. The parameter plane is a record of these behaviors. Mathematicians often use pictures such as these to record information. When you plot the fates of the orbits like this, you begin to see the big picture—the whole story. The famous Mandelbrot set is a picture just like this: It is the parameter plane for the iteration of quadratic iteration rules.

1. Consider the iteration rule $x \to 0.9x + 1$. What is the fixed point for this rule? Use a calculator or computer to compute the orbit of the seed $x_0 = 100$. How many iterations do you need before this orbit is within 0.01 of the fixed point? Within 0.001?
2. Find a linear iteration rule that has an attracting fixed point at $x_0 = 7$.
3. Find a linear iteration rule that has a neutral fixed point at $x_0 = 3$ and a cycle of period 2.
4. Find a linear iteration rule that has a fixed point at $x_0 = 7$ and $x_0 = 365$.
5. Consider the iteration rule $x \to 0.5\,|x|$. The point $x_0 = 0$ is a fixed point. Is it attracting, repelling, or neutral?
6. Consider the iteration rule $x \to 2\,|x|$. The point $x_0 = 0$ is a fixed point. Is it attracting, repelling, or neutral?

TEACHER NOTES

Graphical Iteration

OVERVIEW

This lesson describes how to graphically represent the iteration process by constructing a web diagram. This is perhaps the most important section in the book since we will use these web diagrams throughout the rest of the book to visualize iteration. Students will need ample time to master this graphing process.

MATHEMATICAL PREREQUISITES

Students need to be familiar with linear expressions of the form $Ax + B$, the notion of slope, and the graphs of linear functions.

MATHEMATICAL CONNECTIONS

Connections to the algebra curriculum include **graphing linear equations, identifying slope,** and **solving systems of linear equations** by graphing or substitution. This lesson could be used in a first-year algebra curriculum with the possible exception of the references to the fixed-point equation.

TECHNOLOGY

Although it is much more instructive for students to plot web diagrams by hand at first, using technology to improve the speed and accuracy of generating these web diagrams will free students up to analyze the results. Many modern graphing calculators can perform graphical iteration; consult your manual for directions. You may also use the applet titled linear web on our Web site **http://math.bu.edu/DYSYS/applets**. This applet allows you to see both the time series and the web diagram at the same time. You may also change the parameters A and B interactively and watch how these two images change. This is the type of experimentation students need to see in order to begin

understanding the importance of these parameters in predicting the fate of orbits and the type of fixed points.

SUGGESTED LESSON PLAN

CLASS TIME

Two 50-minute periods are necessary to cover this material completely. The first may be used to introduce graphical iteration (through the section "Summary of graphical iteration"). Be sure to leave ample time for students to practice constructing web diagrams on their own—both in class and for homework. The next class period should cover classifying fixed points according to the value of *A*.

PREPARATION

Read through the explanation. Be sure to have the process of graphical iteration firmly in hand. It is easy to get confused and go in the wrong direction. Remember: Always go vertically to the graph first, then horizontally back to the diagonal. (Doing this in the opposite direction gives orbits for the inverse function and makes attracting fixed points repelling, and so forth.) You might want to prepare a transparency with the graphs of $y = 2x + 1$ and $y = x$ already on it so that you can use it to demonstrate the construction of a web diagram.

LESSON DEVELOPMENT

DAY 1

After describing and demonstrating the basic process of graphical iteration, either manually on transparencies or by using technology, have students work Investigations 1–3 either in class or as homework. We will use this process throughout this book, so students need to master it. All or part of Further Exploration 1 can be used to solidify this notion.

DAY 2

In the second class period, use graphical iteration to classify fixed points as attracting ($|A| < 1$), repelling ($|A| > 1$), or neutral ($|A| = 1$). If students are having difficulty seeing the connection between *A* and the type of fixed points, have them draw a web diagram for several iteration rules from each category of *A* and have them summarize their results and observations.

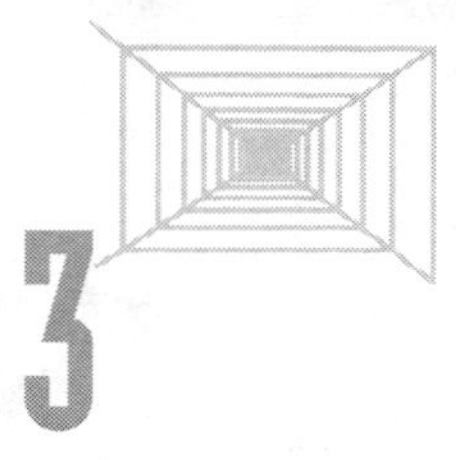

Emphasize the relation between time-series graphs and graphical iteration for various types of fixed points. The applet mentioned earlier can be used to make this connection.

Investigations 4 and 5 are central to our philosophy of having students visualize orbits in different ways: as time series and as web diagrams.

The discussion of neutral fixed points can be left as homework (Investigation 6).

It is important that students be able to summarize the classification of fixed points in words (Investigation 7). This is an excellent example of the necessity of investigating certain mathematical functions or phenomena by separating them into cases such as $|A| < 1$, $|A| > 1$, and $|A| = 1$.

If the parameter plane investigation in Lesson 2 was assigned, have students make a similar classification using pictures (Investigation 8).

LESSON NOTES

Have students compare the time-series graph for $x \to 2x + 1$ to the corresponding web diagram. It is important that they see the relationship between the three representations of the same orbit: web diagram, time-series graph, and numerical iteration table.

Emphasize the roles of A and B as parameters. As these parameters change, we see different phenomena. Sometimes a change in A makes a very big change in the fate of orbits (for example, when A passes through either 1 or –1). Students should have an appreciation for how the choice of A affects the fate of orbits, the type of fixed point, and so forth. This is a prelude to the notion of **bifurcation** that we will encounter in subsequent sections (see Lesson 5, "Nonlinear Iteration").

In the previous lesson, you learned to analyze linear iterations of the form $x \rightarrow Ax + B$ more or less completely. You can determine the different fates of orbits depending on the parameters A and B, and you can decide whether the fixed points are attracting, repelling, or neutral. All of this was done via algebraic methods. This is wonderful when such methods work. Unfortunately, once we move beyond linear iteration, algebraic methods no longer work very well and may even be completely useless. So we need another way to analyze the fate of orbits. This is where geometry comes to the rescue. In this lesson, we will introduce a geometric tool to help us understand iteration. This tool, called **graphical iteration,** will allow us to use the graph of our iteration rule to determine the fate of orbits.

Graphical Iteration

To illustrate graphical iteration, let's begin with the linear iteration rule $x \rightarrow 2x + 1$. We saw earlier that all orbits (except the fixed point) tend to either positive or negative infinity for this iteration rule. We'll now see this geometrically.

The first step in graphical iteration is to draw the graph of $y = 2x + 1$. This graph is a straight line with slope 2 and y-intercept 1. Next, we draw the line $y = x$ on the same graph. The $y = x$ line is a straight line through the origin with slope 1. This line is called the **diagonal**. Graphical iteration will enable us to see the orbit along the diagonal. Here is the picture so far:

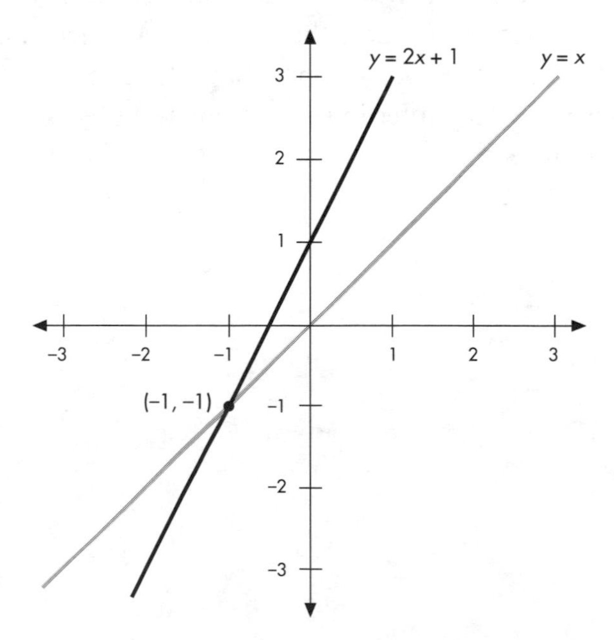

Note that we can "see" the fixed point for this iteration rule already. The fixed point is given by solving the equation $2x + 1 = x$ so that the fixed point occurs at the intersection of the two lines $y = 2x + 1$ and $y = x$. In this case these two lines meet at the point $(-1, -1)$, so our fixed point is -1.

Now suppose you start with the seed $x_0 = 1$ and try to visualize the corresponding orbit on the graph. The next point on the orbit is $x_1 = 3$. Here is how to see this. Locate the seed $x_0 = 1$ on the x-axis. Begin on the diagonal at the point (1, 1), which lies directly over the seed $x_0 = 1$, and then draw a vertical line up to the line $y = 2x + 1$. You reach this graph at the point (1, 3) since $y = 2 \cdot 1 + 1 = 3$. Note that this y-coordinate is exactly the next number on the orbit. Now draw a horizontal line over to the diagonal. This horizontal line is the line $y = 3$, so you reach the diagonal at the point (3, 3). Note that this point lies directly above the next point on the orbit, namely, $x_1 = 3$ on the x-axis. That is, to move from one point on the orbit to the next, we first draw a vertical line to the graph, then a horizontal line back to the diagonal. The numbers corresponding to the orbit appear as the points on the diagonal, as shown here:

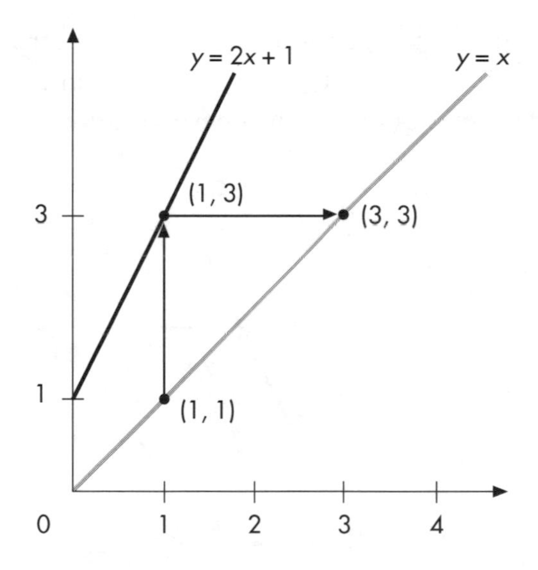

Now keep going. From (3, 3) draw a vertical segment to the line $y = 2x + 1$. This vertical segment has x-coordinate 3, so when you reach $y = 2x + 1$, the y-coordinate is 7. As before, the y-coordinate of this point is the next point on the orbit, $x_2 = 7$. From (3, 7) draw a horizontal segment back to the diagonal.

This is the line $y = 7$. You hit the diagonal at (7, 7), directly over $x_2 = 7$ on the x-axis, as shown here:

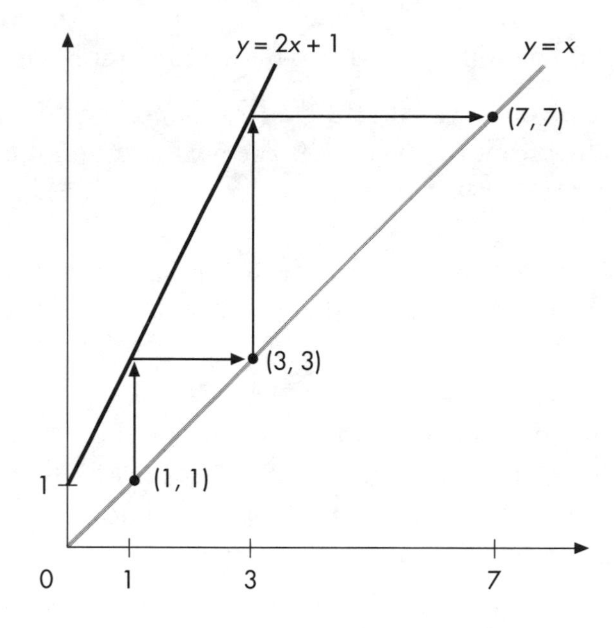

To iterate one more time, draw a vertical segment from (7, 7) on the diagonal to the line $y = 2x + 1$, reaching it at (7, 15). Then draw a horizontal segment to the diagonal, reaching (15, 15). You have found $x_3 = 15$.

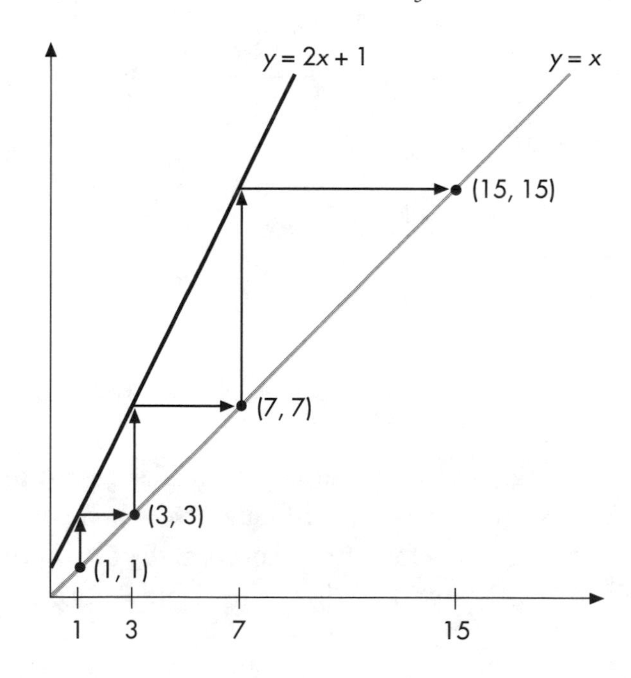

Observe that this process yields a "staircase." Each step in the staircase ends directly *over* the next point of the orbit, so we say that each step in the staircase leads you *to* the next point on the orbit. You see immediately that, if you continue iterating forever, the staircase leads to infinity. Of course, that is exactly what is happening to the orbit! This is another way of seeing the fate of the orbit of $x_0 = 1$.

THE TIME-SERIES GRAPH

Before leaving this example, let's look at the associated time-series graph, which gives a very different picture of this orbit. The iteration rule is $x \longrightarrow 2x + 1$, and the seed is $x_0 = 1$. We know that the orbit begins

$$1 \longrightarrow 3 \longrightarrow 7 \longrightarrow 15 \longrightarrow \cdots$$

So here is the time series:

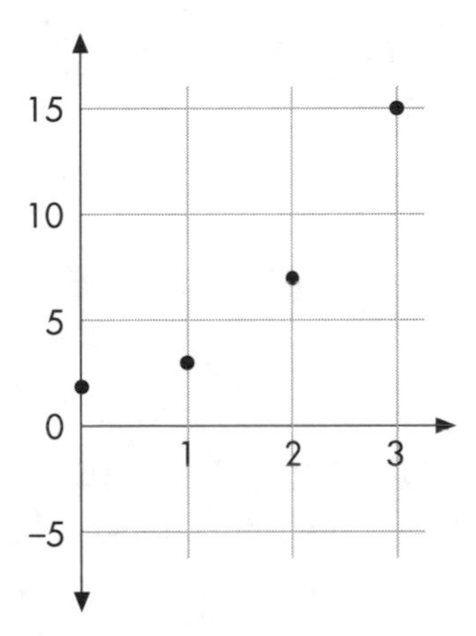

Note how different this image is from the graphical iteration. But both pictures convey the same information: The orbit is increasing toward infinity. You will see that sometimes the time-series graph is a more helpful illustration of the fate of the orbit, but other times the graphical iteration is better.

OTHER ITERATION RULES

Although graphical iteration leads to a staircase graph for the iteration rule $x \rightarrow 2x + 1$, this is not always the case. For example, consider the iteration rule $x \rightarrow -x/2$ with seed $x_0 = -2$. The orbit is

$$-2 \rightarrow 1 \rightarrow -\frac{1}{2} \rightarrow \frac{1}{4} \rightarrow -\frac{1}{8} \rightarrow \cdots$$

which tends to the attracting fixed point at 0.

Here is how graphical iteration displays this orbit. Note how we begin on the diagonal, then proceed vertically to the graph, and then horizontally to the diagonal, just as before:

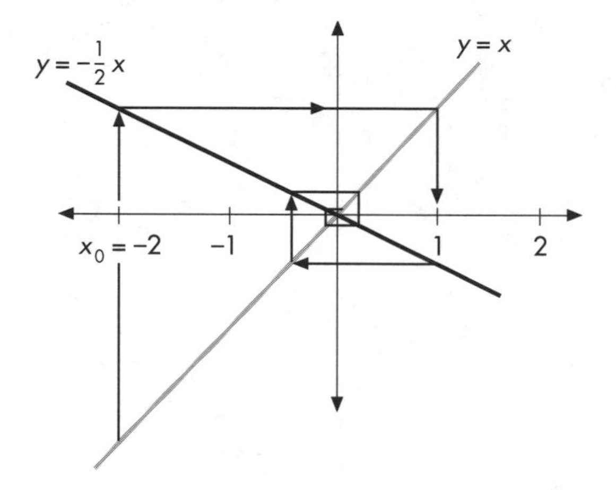

Rather than climbing or descending a staircase, this graphical iteration hops back and forth on either side of 0. This graphical iteration is sometimes called a **web diagram,** because it resembles a spider's web.

As another example, the iteration rule $x \to -x + 1$ with seed $x_0 = -\frac{1}{2}$ yields a different picture:

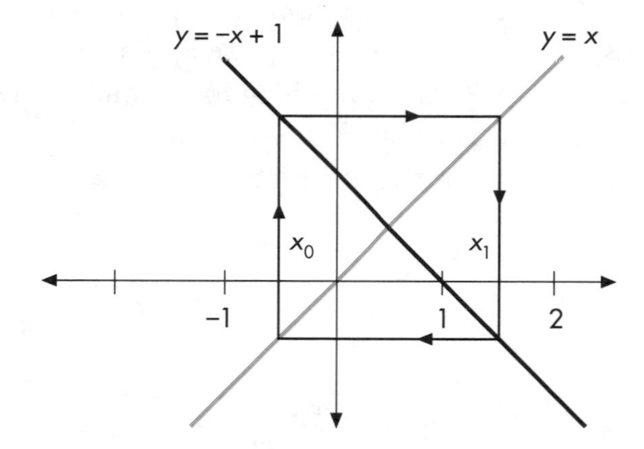

Here the orbit of x_0 lies on a 2-cycle, which is represented in graphical iteration as a square.

SUMMARY OF GRAPHICAL ITERATION

To view the orbit of seed x_0 under the linear iteration rule $x \to Ax + B$, you always perform the following steps:

1. Graph both the lines $y = Ax + B$ and $y = x$.
2. Start at (x_0, x_0) directly over (or under) the seed and draw the vertical segment to $y = Ax + B$. You reach the graph at $(x_0, Ax_0 + B)$.
3. Now draw the horizontal segment from $(x_0, Ax_0 + B)$ to the diagonal line $y = x$, landing on the diagonal at the point $(Ax_0 + B, Ax_0 + B)$.
4. Then repeat steps 2 and 3, using $x_1 = Ax_0 + B$ as the starting value to locate x_2, and so on.

The important thing to remember is that you **always move vertically to the graph first, then horizontally to the diagonal.**

In the previous lesson, we found fixed points for the linear iteration rule $x \to Ax + B$ for various values of A and B. We saw that these points could be attracting $(|A| < 1)$, repelling $(|A| > 1)$, or neutral $(|A| = 1)$. Now let's see this same fact using graphical iteration.

SEEING A REPELLING FIXED POINT

When $A > 1$, the graph of $y = Ax + B$ crosses the diagonal at the fixed point as shown below. The slope of the graph is steeper than that of the diagonal. This means that seeds that are close to the fixed point have orbits that move away from the fixed point. This is a **repelling** fixed point. Moreover, from graphical iteration, we see that not only do these orbits move away from the fixed point, but they also tend to positive or negative infinity.

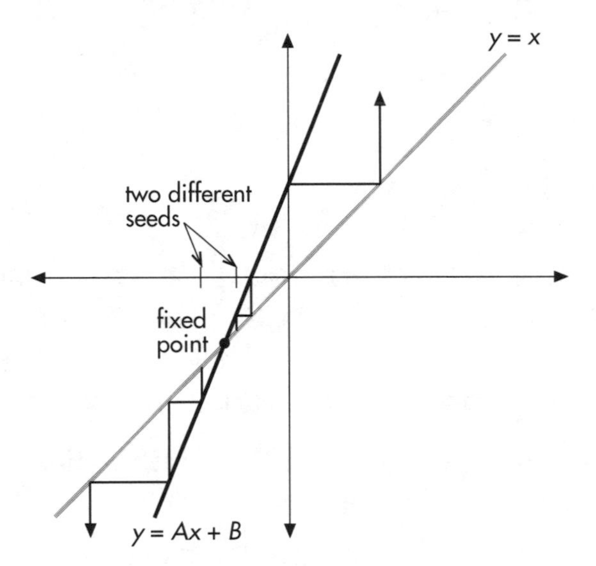

If the slope of $y = Ax + B$ is negative and steep ($A < -1$), a similar phenomenon occurs: The graph crosses the diagonal at a fixed point as before, but now orbits hop from left to right and back again as they move away from this fixed point, as shown below. This kind of fixed point is also called a repelling fixed point, since orbits tend to move away from it.

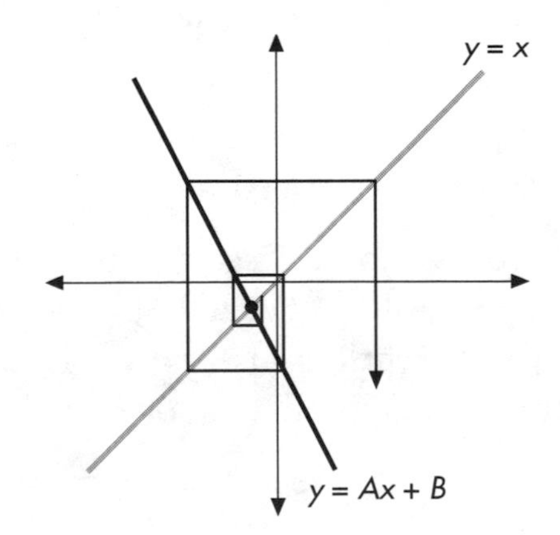

SEEING AN ATTRACTING FIXED POINT

When $-1 < A < 1$, the situation is completely different. Now the slope of $y = Ax + B$ is less steep than the diagonal. This forces nearby orbits to tend toward the fixed point, so we have an **attracting** fixed point.

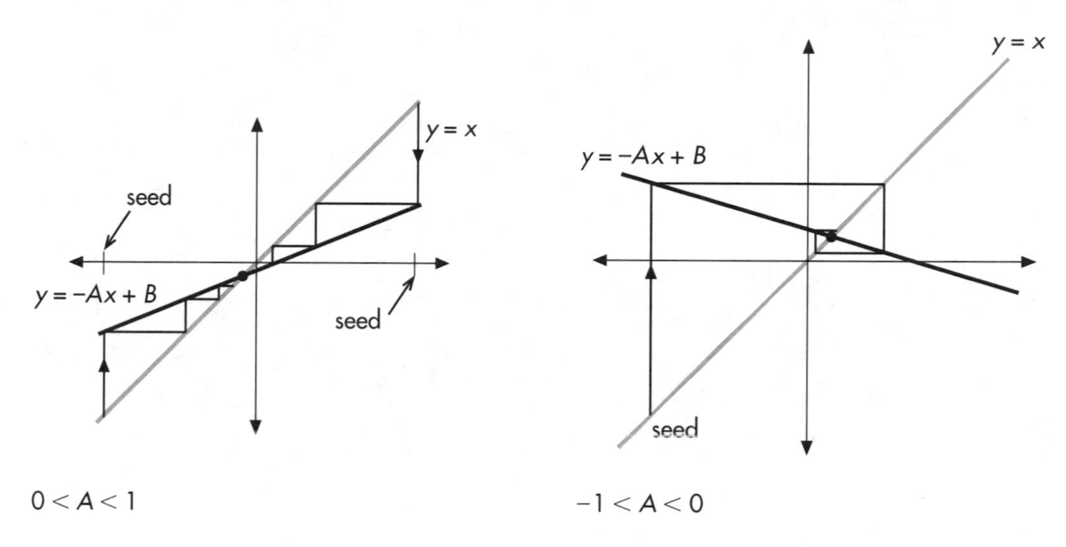

There is one special case of an attracting fixed point, when $A = 0$. In this case, the graph of our iteration rule is a horizontal line $y = B$. Thus, there is one fixed point, at $x_0 = B$, and all other orbits are eventually fixed since they land on B after one iteration.

1 ▹ GRAPHICAL ITERATION AND TIME SERIES

Sketch the first five iterations for the iteration rule $x \to 2x + 1$ and for the indicated seeds. Then draw the associated time series.

a. $x_0 = 0$

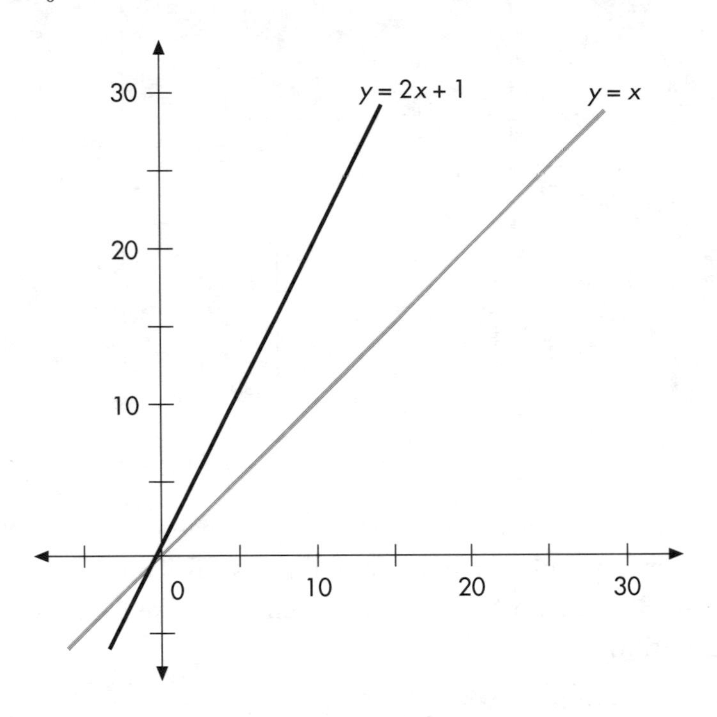

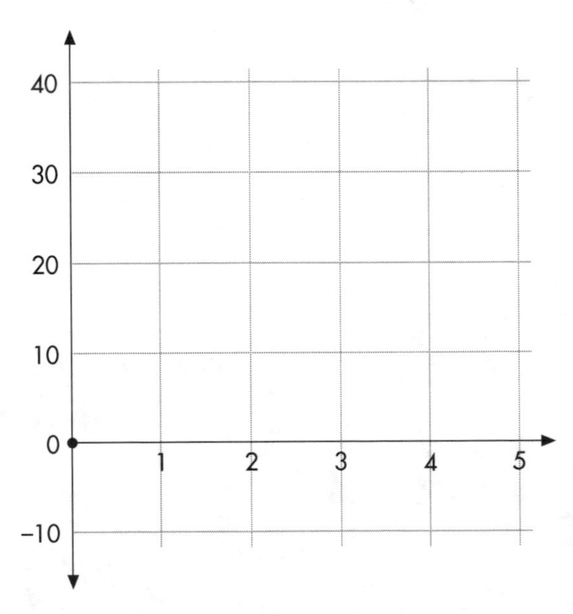

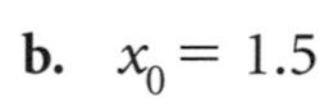

b. $x_0 = 1.5$

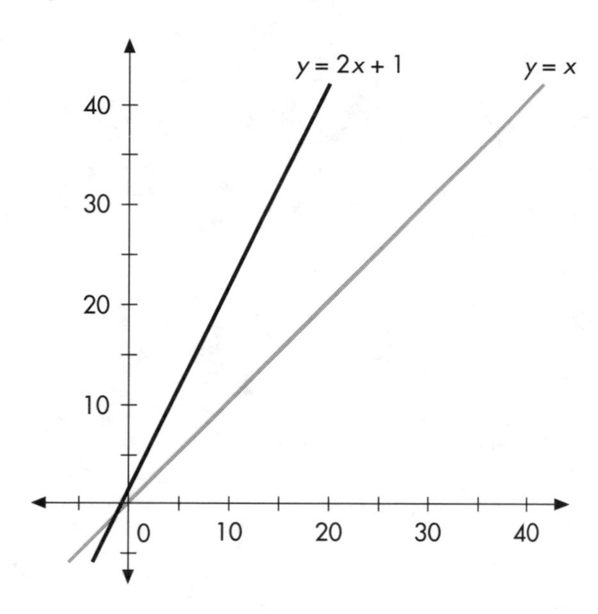

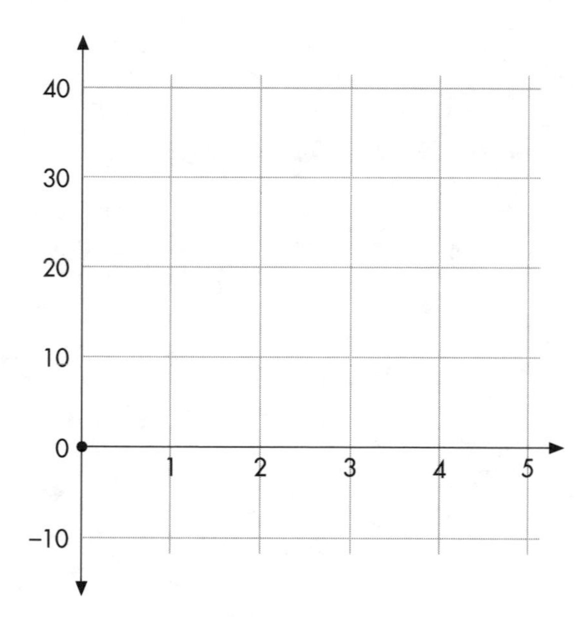

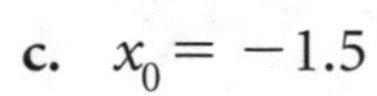
c. $x_0 = -1.5$

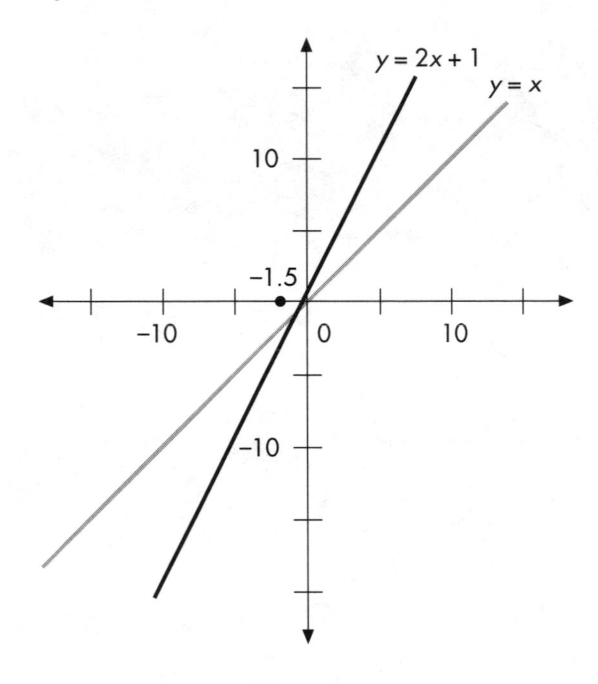

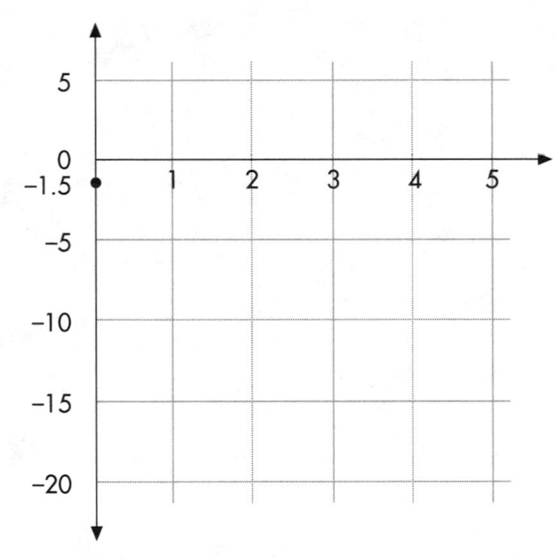

2 ▹ FINDING THE FATE OF THE ORBIT GEOMETRICALLY

Here are the graphs of several linear iteration rules. In each case, use graphical iteration to sketch the fate of the orbit of the seed x_0. We have not specified the numerical value of x_0; you should work strictly geometrically. Remember: Always start on the diagonal and first go vertically to the graph of the iteration rule, then horizontally back to the diagonal. When you are finished, describe the fate of the orbit of x_0.

a.

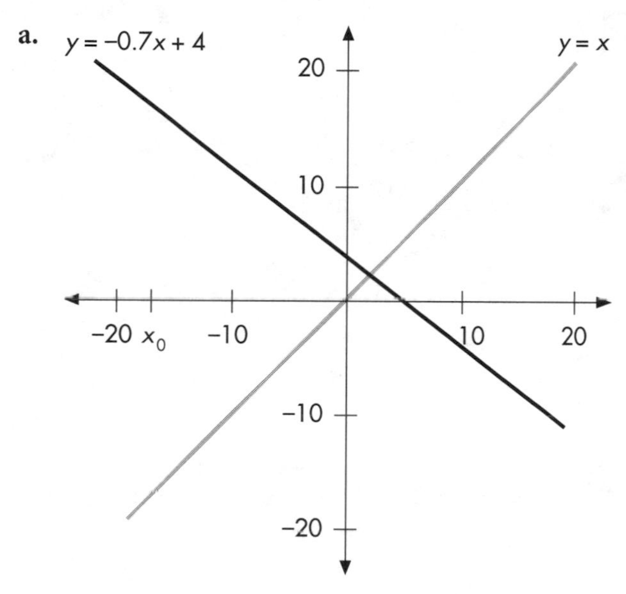

Fate of the orbit of x_0: ______________________

b.

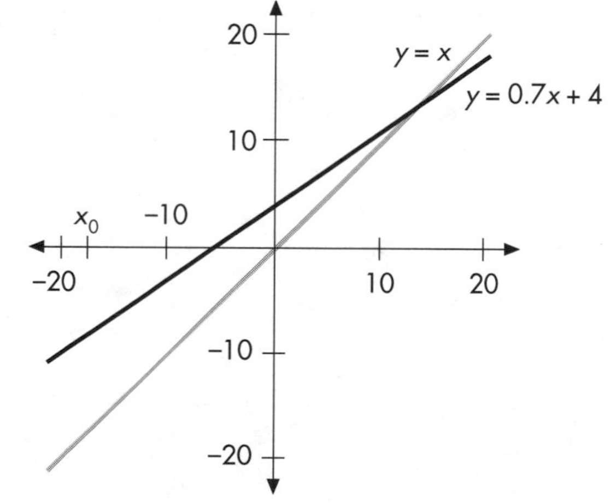

Fate of the orbit of x_0: ______________________

3 ▹ Graphical iteration

For each of these linear iteration rules, first sketch the graph of the given rule. Then use graphical iteration to display the fate of the orbit of the given seed.

a. $x \rightarrow 3x - 1$, $x_0 = 0$

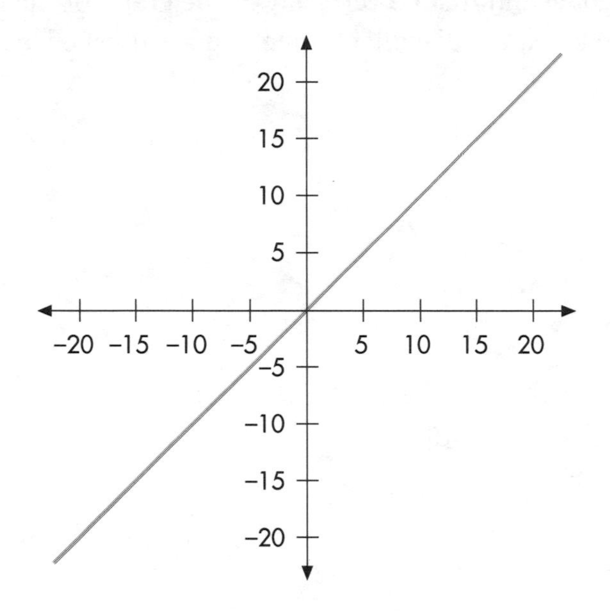

b. $x \rightarrow \frac{x}{2} + 2$, $x_0 = -20$

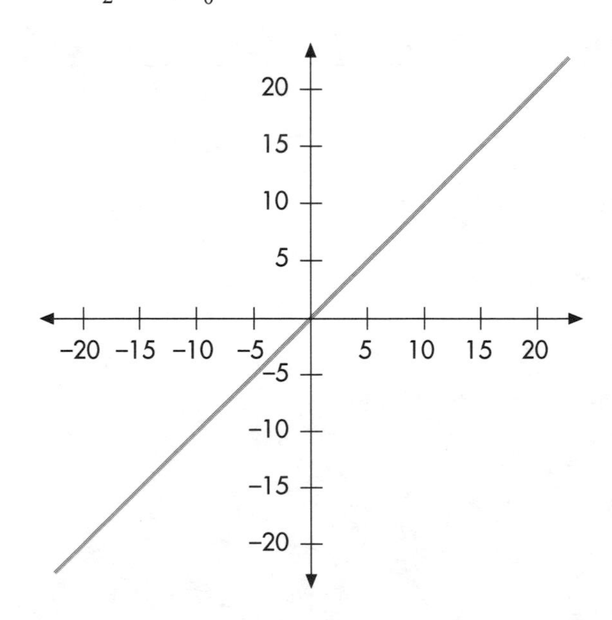

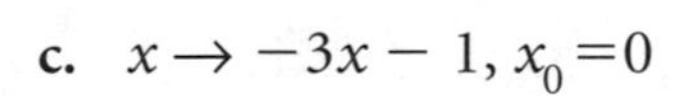
c. $x \to -3x - 1, x_0 = 0$

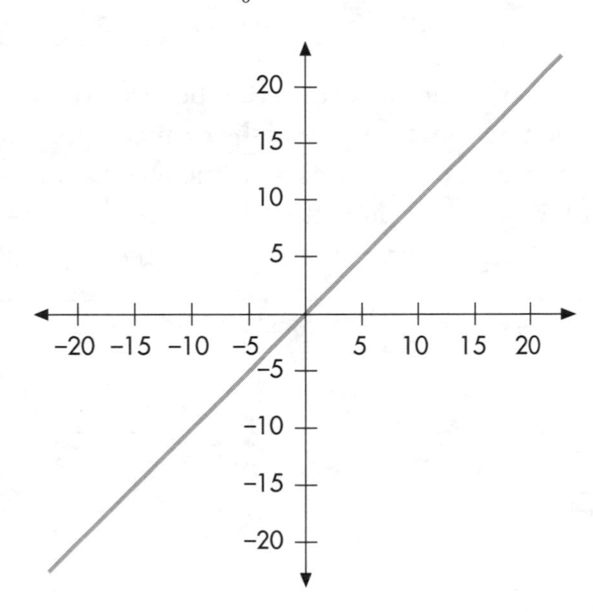

4 ▹ MATCHING ORBITS, TIME SERIES, AND GRAPHICAL ITERATION

In order to make sure you understand the connection between orbits, time-series graphs, and graphical iteration, try to make the connections you need to complete the table. Here are pieces of six orbits, six time series, and six graphs. Determine which goes with which. Then identify the linear iteration $x \rightarrow Ax + B$ that generated each. Lastly, sketch the graphical iteration on each graph.

	Orbit	Time series	Graphical iteration	$x \rightarrow Ax + B$
a.	0, 2, 6, 14			
b.	$0, -6, -8, -\frac{26}{3}$			
c.	0, 1, 2, 3			
d.	$0, -6, -4, -\frac{14}{3}$			
e.	$0, -4, 0, -4$			
f.	0, 3, 3, 3			

A.
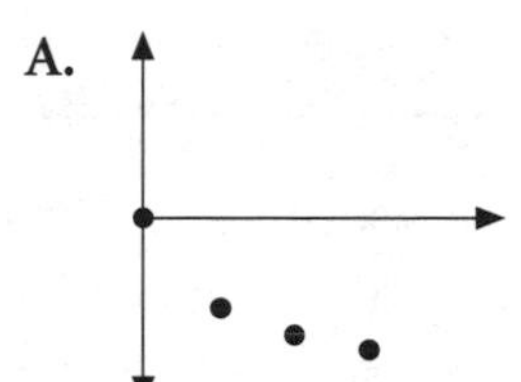

B.
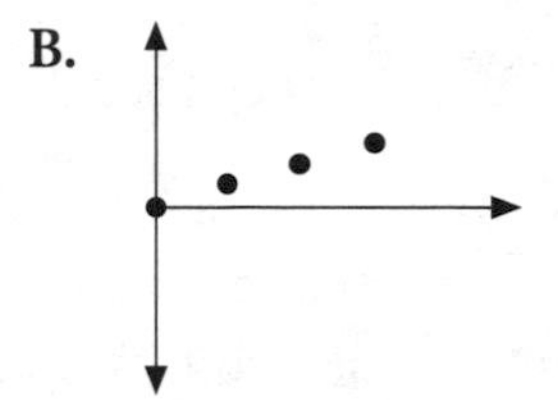

C.
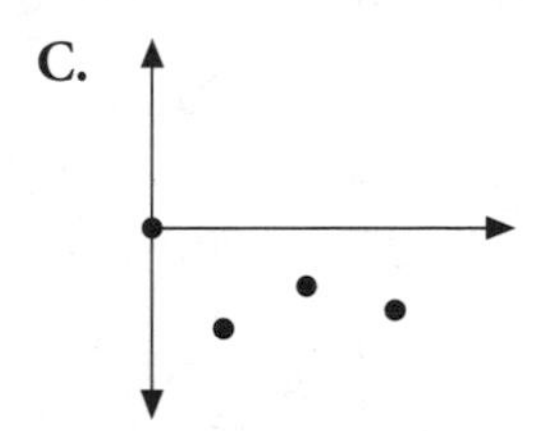

D.
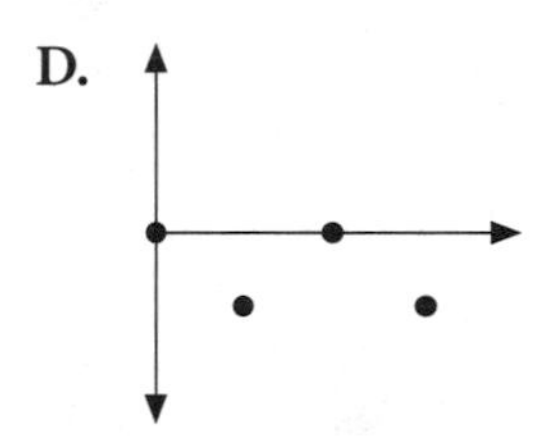

E.
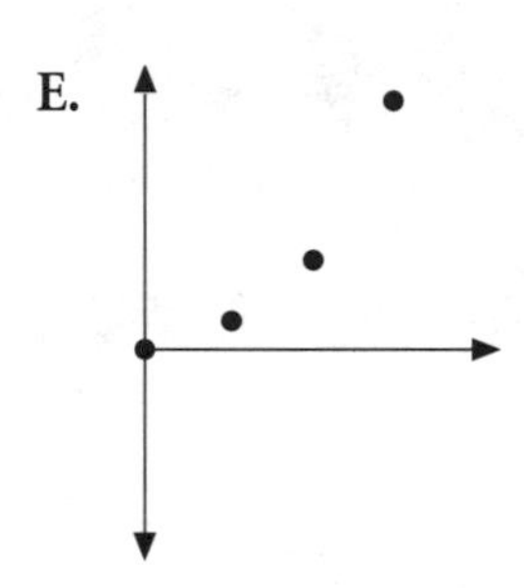

F.
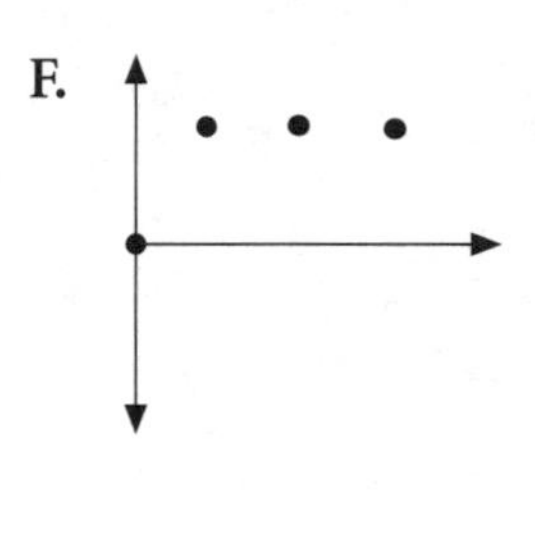

I.
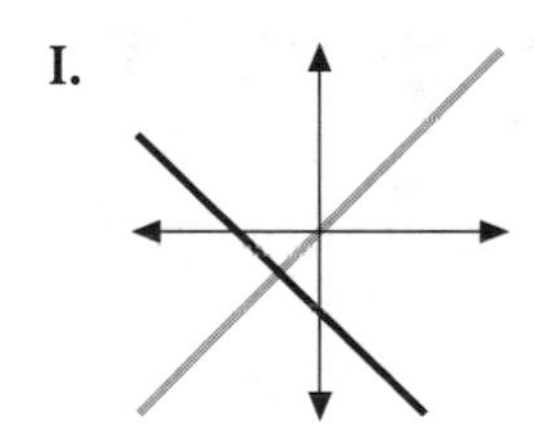

II.
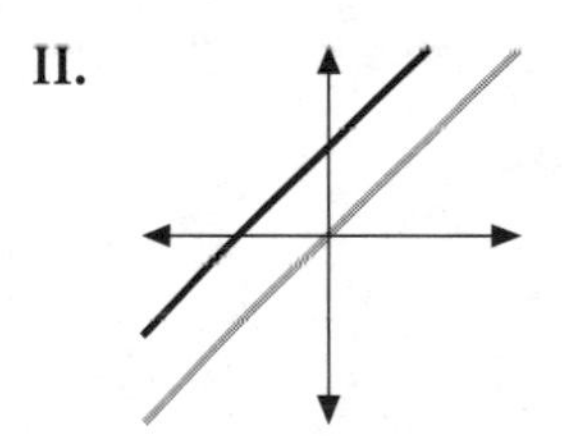

III.
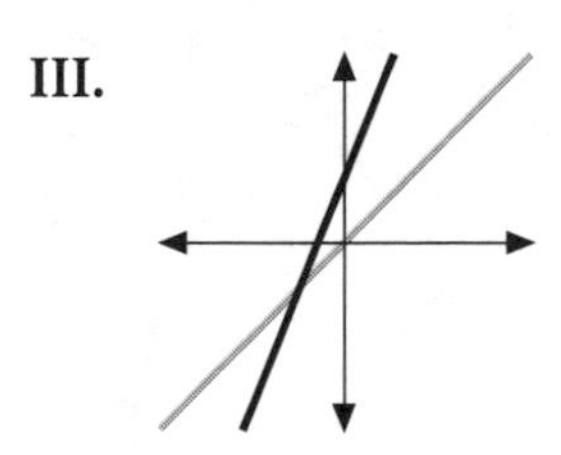

IV.
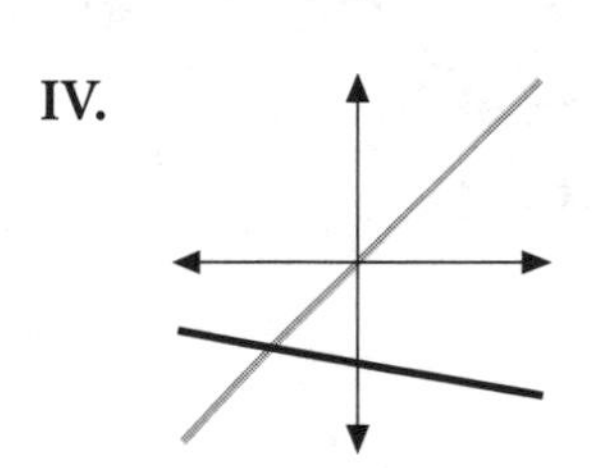

V.
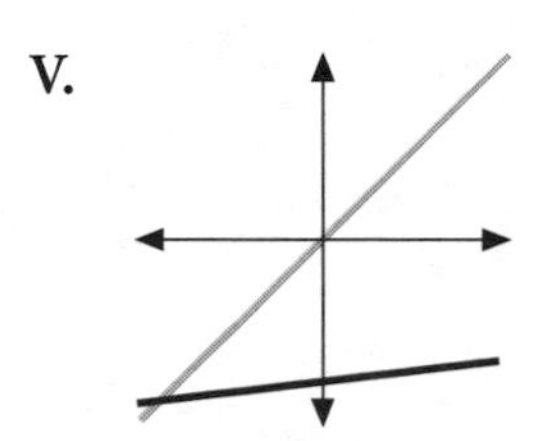

VI.
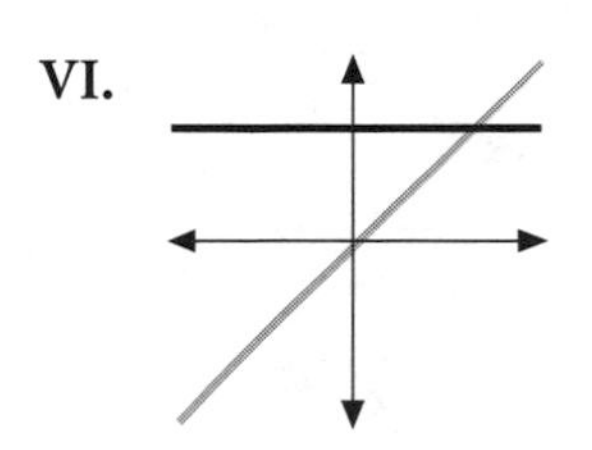

5 ▹ Geometric matching

Try the match from Investigation 4 again, this time given seven time series and six graphs. Show which goes with which. Note that this time we haven't given you the numbers on the orbit: You have to work geometrically. List the matches here: ____

1.

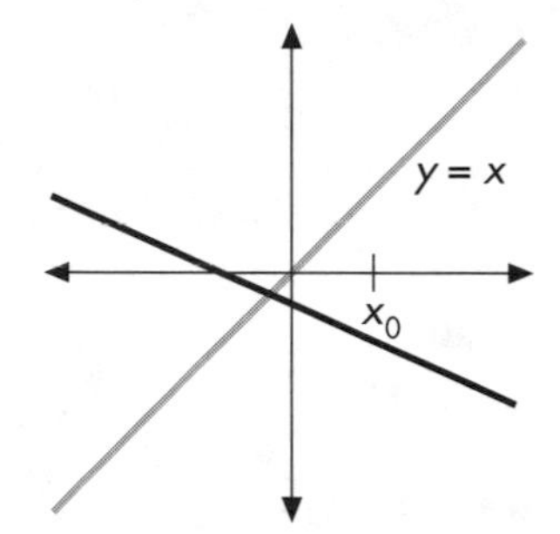

A.

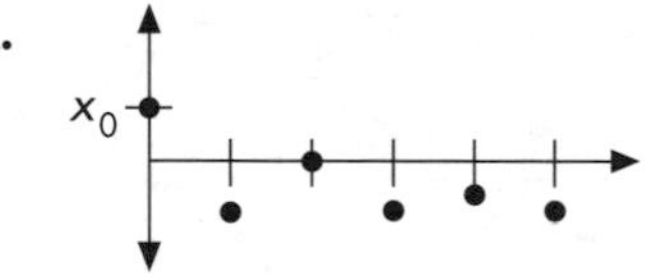

2.

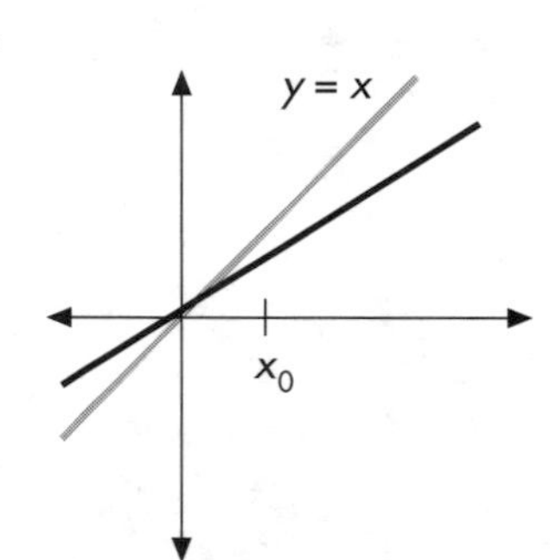

B.

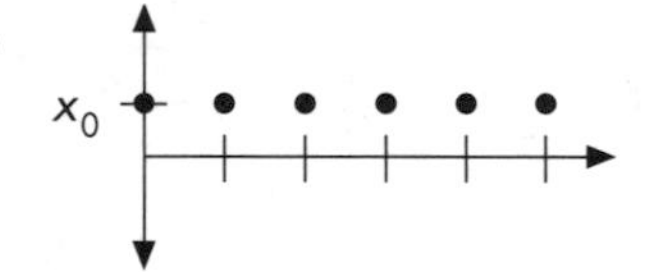

3.

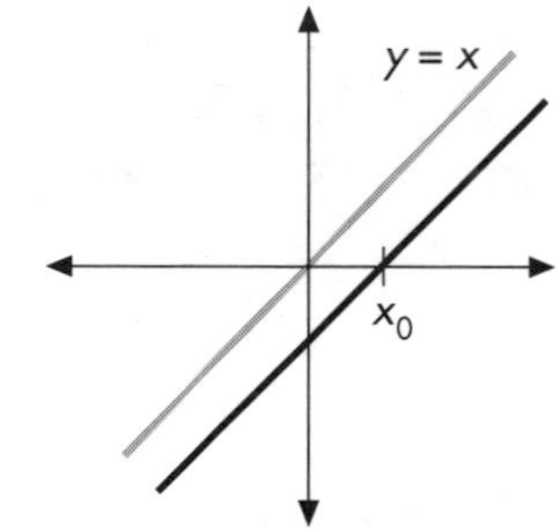

C.

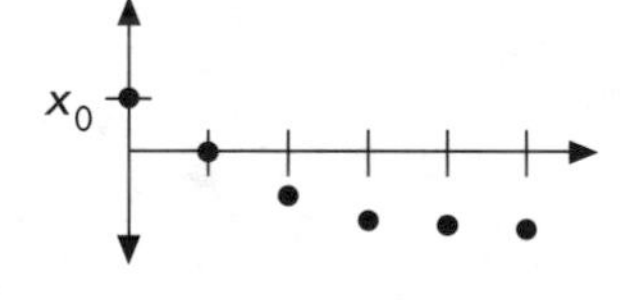

4.

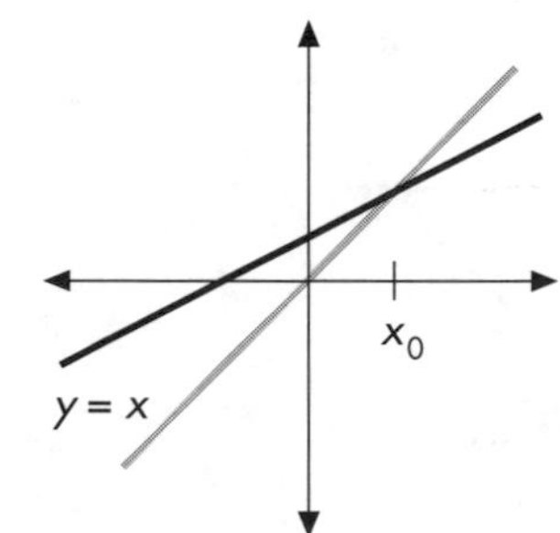

D.

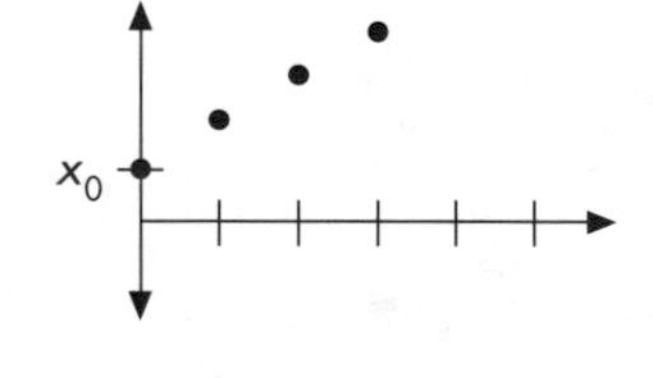

5.

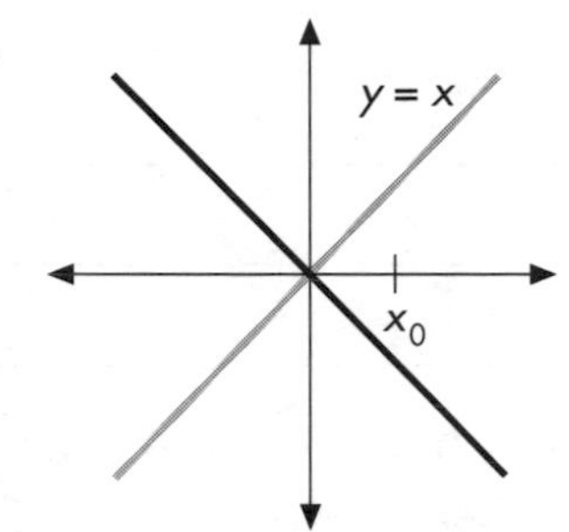

E.

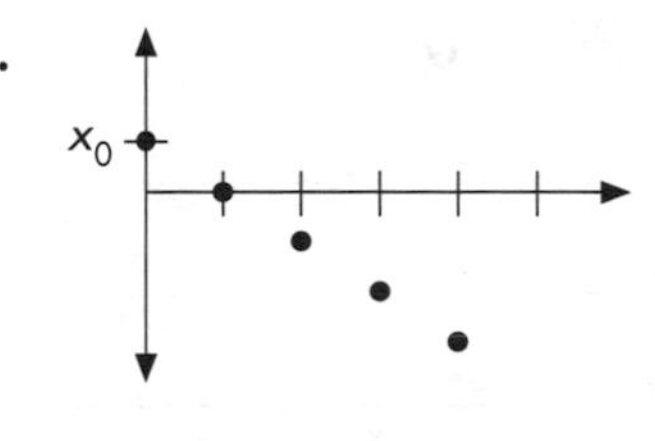

6.

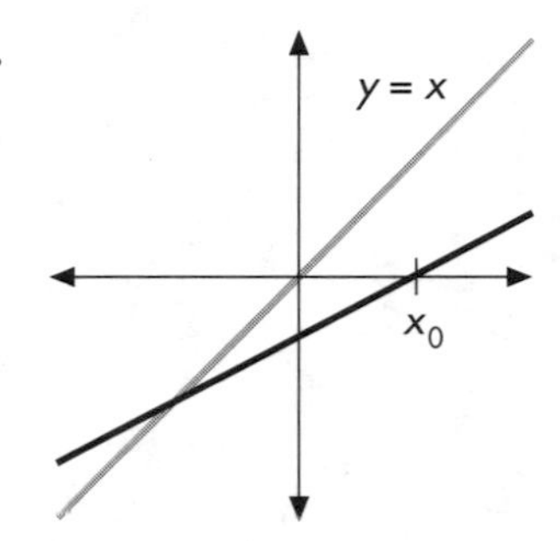

F.

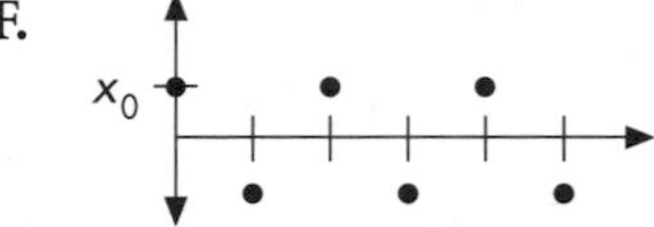

G.

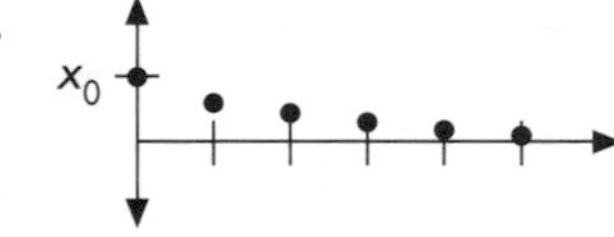

6 ▸ Neutral fixed points

Remember: Neutral fixed points occur when $A = 1$ or $A = -1$. Using graphical iteration, sketch what happens to the orbits of $x \rightarrow -x + B$.

a. First choose a value of B. Then draw the graph of the iteration rule and use graphical iteration to plot several orbits. Then discuss the fate of the orbits on the next page.

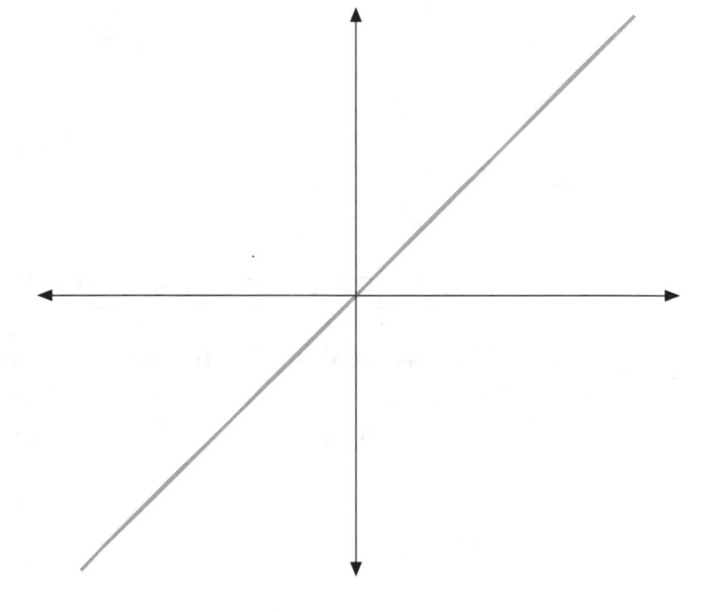

b. Fate of the orbits:

$x_0 =$ ______________________________

$x_0 =$ ______________________________

$x_0 =$ ______________________________

c. Discuss the situation that occurs when $A = 1$ but $B \neq 0$. First sketch the graphical iteration, then discuss the fate of the orbits below. Discuss both the cases $B > 0$ and $B < 0$.

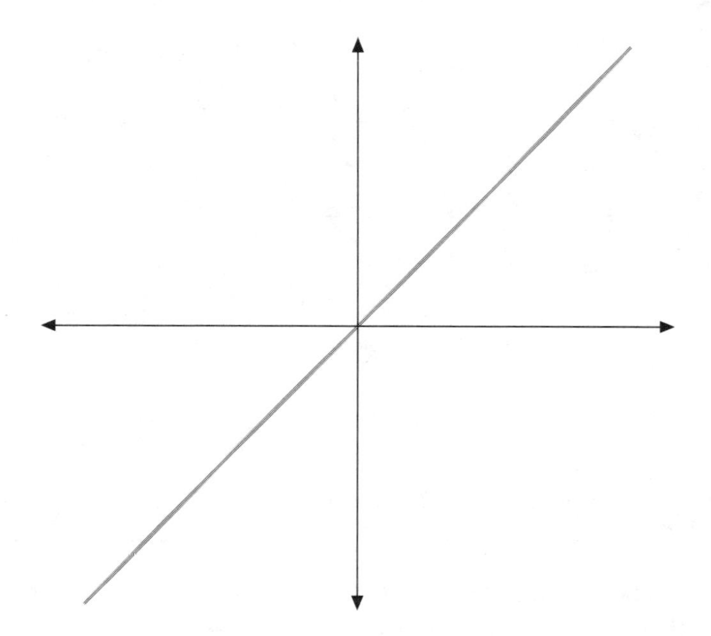

d. Fate of the orbits:

$x_0 =$ ______________________________

$x_0 =$ ______________________________

$x_0 =$ ______________________________

7 ▹ BACK TO THE PARAMETER PLANE

Now we are done. We have determined the fate of all orbits of the linear iteration rule $x \rightarrow Ax + B$ for all possible values of A and B. And we have also classified the fixed points for these rules as attracting, repelling, or neutral. Since this has worked out so successfully, let's summarize what we have found. As we often do in mathematics, we summarize this both in words and in pictures.

Write a sentence below that describes the fate of all orbits for $x \to Ax + B$ in all cases. Don't forget to describe the fixed point (if there is one) and to include all special subcases.

Case **Fate of all orbits**

$|A| > 1$ ____

$|A| < 1$ ____

$A = -1$ ____

$A = 1$ ____

8 ▸ BACK TO THE PARAMETER PLANE

Now let's repeat the same summary, but using geometry this time. We will again paint the picture of the parameter plane for the linear iteration rule $x \rightarrow Ax + B$. This time, however, we will draw the "typical" graphical iteration that occurs in each region. In the *AB*-plane below, draw the regions in which we have different behaviors. In each region, sketch a small representative graphical iteration. Don't forget the special cases where we have a neutral fixed point.

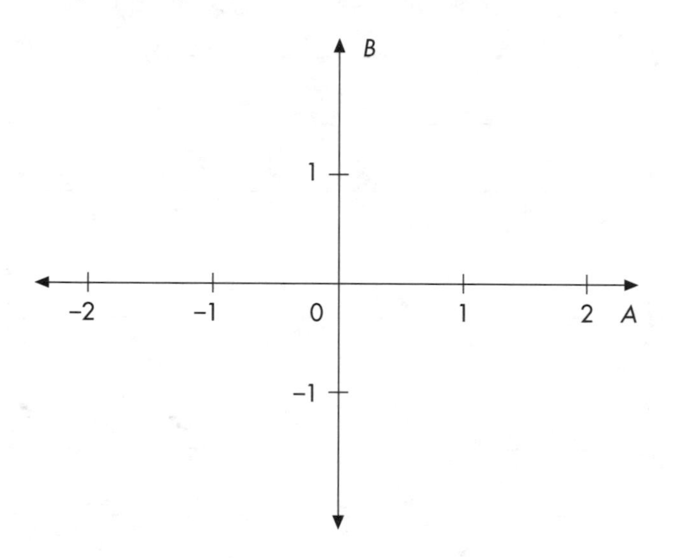

9 ▸ ORBITS FOR LINEAR ITERATION RULES

We have been dealing with some very special iteration rules in this lesson. The orbits that arise from these rules are very simple. Not all orbits come from linear iteration rules. Which of the following pieces of orbits could be orbits for a linear iteration? If the orbit does come from a linear iteration rule, give the rule. If the piece of orbit cannot come from a linear iteration rule, explain why not (perhaps using graphical iteration).

a. 0, 2, 1, 0, 1, 2, . . . ______________________________

b. 8, 4, 2, 1, . . . ______________________________

c. −3, −3, −3, −3, . . . ______________________________

d. 6, 4, 2, 0, . . . ______________________________

e. 6, 4, 2, 10, . . . ______________________________

f. −3, 5, −3, 5, . . . ______________________________

1. For each of these iteration rules, first find the fixed points. Then determine if the fixed point is attracting, repelling, or neutral. Finally, sketch the graphical iteration using a seed of your choice.

   a. $x \rightarrow -7.2x - 1$

   b. $x \rightarrow -0.72x - 1$

   c. $x \rightarrow -720x - 1$

   d. $x \rightarrow -\frac{1}{72}x - 1$

   e. $x \rightarrow -\frac{1}{72}x - x$

   f. $x \rightarrow 7x - 72$

2. Consider the iteration rule $x \rightarrow |x| - 1$. First draw the graph of this expression, then use graphical iteration to indicate the fate of various orbits. Explain in a paragraph or two what you observe.

3. Consider the "doubling function," which is defined on the interval $0 \leq x \leq 1$ by

$$x \rightarrow \begin{cases} 2x & x < \frac{1}{2} \\ 2x - 1 & x \geq \frac{1}{2} \end{cases}$$

   That is, this iteration rule doubles x if $x < ½$. Otherwise, we double x and subtract 1. Remember that $0 \leq x \leq 1$. First compute the orbits of the following seeds. Then draw the graph of this rule. Finally, use graphical iteration to sketch the orbits of each of the following seeds.

   a. $x_0 = \frac{1}{3}$

   b. $x_0 = \frac{1}{7}$

   c. $x_0 = \frac{1}{8}$

   d. $x_0 = \frac{3}{7}$

   e. $x_0 = \frac{1}{9}$

TEACHER NOTES

Measuring Population Growth

Overview

In this lesson, we introduce an application of nonlinear iteration to ecology, namely, the logistic model of population growth. As such, this lesson is optional, though it does serve as motivation for all that comes later since it provides a real-world model of a nonlinear iteration rule. Most of the remainder of this book deals with quadratic iteration rules—the types of rules that arise from the logistic model. The Investigations provide a series of experiments that foreshadow what we will discover for quadratic iteration rules in the following lessons.

Mathematical Prerequisites

Students need to be familiar with linear and quadratic equations and the concept of proportionality. Students who have worked with the applications in *Iteration* (savings plans, loans, and so forth) should have no problem understanding the exponential growth model for population growth.

Mathematical Connections

This lesson has a strong connection to topics in the second-year algebra curriculum such as **direct variation, quadratic functions,** and **exponential growth.** It also provides an opportunity to spend time discussing **mathematical models.**

Technology

Technology is not necessary for introducing the material in this lesson, but a calculator or spreadsheet is essential for completing the Investigations.

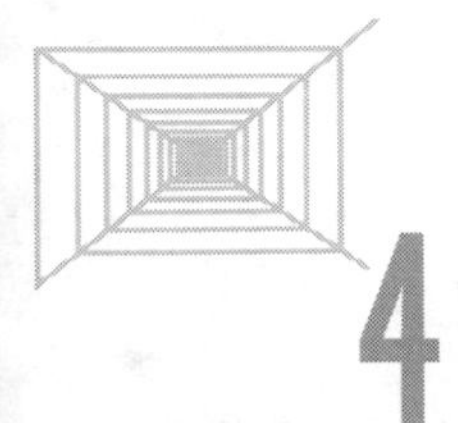

SUGGESTED LESSON PLAN

CLASS TIME

One 50-minute period. The introductory activities on the linear and exponential growth models should be fairly straightforward, leaving the majority of the class period for explanation of the logistic model.

PREPARATION

Read the discussion of how the logistic model is developed. Be aware that there are many other possible models for population growth that fit our assumptions. We have chosen the "simplest" model here. You might also want to talk to some of the science teachers to see if you can connect the class discussion to their curriculum. It is also suggested that you read through Investigation 7 and determine if you want to assign it. If so, you may want to discuss what is meant by not recording "transient" behavior.

LESSON DEVELOPMENT

Begin by discussing the linear population growth model. Since this rule is linear, students should be able to predict the fate of orbits (that is, the fate of the population) without difficulty, given the previous three lessons. After developing the linear model, you might ask students why this is not a realistic model for population growth.

Move on to a discussion of the logistic model. Emphasize the starting assumptions: there is some absolute maximum population L; if P is very small, it means the population grows exponentially; P larger than L means the population goes extinct. You might want to point out to students that the logistic model is quadratic by multiplying the expression out to the form $kx - kx^2$. This is an excellent place to discuss what other features might have been included to make this model more realistic. Investigation 2 is aimed at this topic.

Assign selected problems from Investigations 1–6 depending on whether or not you included the logistic model in the class discussion. See the Lesson Notes below for suggestions.

LESSON NOTES

You could incorporate Investigations 1 and 2 into the class discussion. In Investigation 3, students are asked to find fixed points for the logistic iteration rule. We cover this in detail in the next lesson.

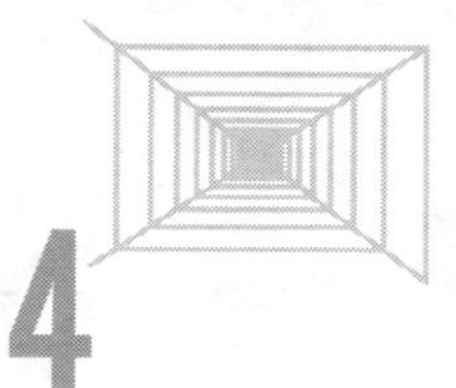

Investigations 3–6 are precursors to what we will do in the next lessons. If you plan to cover this later material in detail, you might wish to postpone assigning these exercises until later, or you might consider having students do these Investigations to help prepare them for the upcoming material. (Basically, students are asked to observe experimentally what happens if the parameter k is varied in the logistic model.)

Investigation 7 foreshadows what will be covered in Lesson 9. If you are not covering that material, this Investigation can be used as a way to have students see this image experimentally without going into all the details. Alternatively, this and the subsequent Investigations provide good material for long-term class projects or mathematical term papers.

One area of science that has benefited immensely from the ideas involving iteration is ecology. Ecologists are often interested in populations of various species. They ask questions like: Will the population of a certain species die out? Is the population tending toward overcrowding? To answer these questions and make predictions about certain populations, ecologists often construct mathematical models that describe how the populations change. Since the populations are often measured only yearly, or in distinct generations, the data that ecologists use can be plotted geometrically as a time-series graph. The scientist then searches for an iteration rule that generates the time series. If such an iteration rule can be found, the ecologist can use it to make predictions about the future.

In this lesson, we introduce several models for population growth. The first is a linear iteration rule that is not very realistic as a population growth model. Nevertheless, we will be able to understand this rule completely using techniques from the previous lessons. The second model, however, is our first example of a nonlinear iteration rule. We will see that understanding this iteration rule is much more difficult. Indeed, determining the fate of all orbits under this iteration rule is still an unsolved problem in mathematics.

THE EXPONENTIAL GROWTH MODEL

In the simplest population growth model, we assume that the population of a certain species in the next generation is directly proportional to the population present in this generation. This makes sense: If we have more of a certain species present, then we expect more babies to be born and so the population in the next generation will be larger. If the population in the present generation is smaller, fewer babies will be born and the population in the next generation will be correspondingly smaller.

Thus, the rule for population growth tells us something about the population in the succeeding generation given information about the population now. So let's say that the population in generation n is P_n. Then our assumption about the population growth is that P_{n+1} is directly proportional to P_n so that $P_{n+1} = kP_n$ where k is the constant of proportionality. That is, this model for population growth is governed by the linear iteration rule $x \rightarrow kx$. This is called the **exponential growth model** for population growth.

For example, if we know that the initial population P_0 is 1000 and that, in the next generation, the population P_1 is 1100, then we can predict the future of this population using the exponential growth model as follows. With $P_0 = 1000$

and $P_1 = 1100$, we must have $P_1 = kP_0$ or $1100 = k \cdot 1000$. Therefore, we have $k = 11/10$. Thus, this particular iteration rule is $x \rightarrow 11x/10$.

Using this rule, we then compute the orbit of 1000 as usual:

$$1000 \rightarrow 1100 \rightarrow 1210 \rightarrow 1331 \rightarrow 1464.1 \rightarrow \cdots$$

This last number needs a little explanation. We do not mean to imply that in the fourth generation there is a population of exactly 1464.1. (It is interesting to think about a tenth of a person or a tenth of a moose.) Rather, our predictions here are purely mathematical and, in fact, only approximate. Also, in many cases we measure populations not in actual numbers of individuals present but rather, in other, more appropriate units. For example, populations of people are often measured in millions of people, or populations of moose in thousands. In these cases, one-tenth of a unit simply means one-tenth of a million or one-tenth of a thousand.

Analyzing the linear iteration rule $x \rightarrow kx$ is straightforward using the methods of the preceding lessons. There is only one fixed point for this rule (usually), and that fixed point is located at 0. This too makes sense: If there is no population present in this generation, there are no babies born and so we do not expect any population in the next generation either. Since populations are positive numbers, we are interested in the fate of orbits under this rule only when both the seed and the constant k are positive.

For example, suppose that $k = 2$. That is, our iteration rule is $x \rightarrow 2x$. The orbit of x_0 is given by

$$x_1 = 2x_0$$

$$x_2 = 2x_1 = 2 \cdot 2x_0 = 2^2x_0$$

$$x_3 = 2x_2 = 2 \cdot 4x_0 = 2^3x_0$$

and the population in the nth generation is

$$x_n = 2^n x_0$$

That is, as long as $x_0 > 0$, the population doubles each year and we have a population explosion. For example, if the initial population is $x_0 = 10$, the orbit is

$$10 \rightarrow 20 \rightarrow 40 \rightarrow 80 \rightarrow 160 \rightarrow 320 \rightarrow 640 \rightarrow \cdots$$

Note also that we have a repelling fixed point at 0.

If, on the other hand, $k = 1/2$, then our population tends to extinction. This happens because we have

$$x_1 = \frac{x_0}{2}$$

$$x_2 = \frac{x_1}{2} = \frac{x_0}{4}$$

$$x_3 = \frac{x_2}{2} = \frac{x_0}{8}$$

and, in general,

$$x_n = \frac{x_0}{2^n}$$

So we see that the population is halved each generation and tends to 0. For example, if the initial seed x_0 is 100, the orbit is

$$100 \rightarrow 50 \rightarrow 25 \rightarrow 12.5 \rightarrow 6.25 \rightarrow 3.125 \rightarrow 1.5625 \rightarrow \cdots$$

The population decreases and we have an attracting fixed point at 0 and all orbits tend to it.

Obviously, the exponential growth model is quite simplistic. To account for other factors that affect population growth, we need to modify the model.

THE LOGISTIC POPULATION GROWTH MODEL

One aspect of population growth that the exponential growth model neglects is overcrowding. If the population of a certain species becomes too large, there will not be enough food for individuals to survive, famine may set in, and the population may decline or even become extinct. Let's try to create a model that takes into account the effects of overcrowding.

One way to do this is to make two assumptions. The first is that as long as the population is relatively small, the population will grow more or less as in the exponential growth model. That is, as long as P_n is small, P_{n+1} will be directly proportional to P_n. But when P_n is large, something else happens. Let's assume that there is some absolute maximum population, say L, at which disaster strikes. If the population ever reaches or exceeds L in a given generation, in the next generation, the species becomes extinct. You might think of L as being such a large population that every square inch of space is used up by the species. There is no room to grow food, no room to move around, and so the population becomes extinct.

There are many ways to take these two assumptions into account. One of the easiest is to assume that

$$P_{n+1} = kP_n(L - P_n)$$

where k is some positive constant of proportionality. Notice that if P_n is small, the term $L - P_n$ is close to L, so P_{n+1} is essentially given by kLP_n. That is, P_{n+1} is again directly proportional to P_n, just as in the exponential growth model.

On the other hand, if $P_n = L$, then $P_{n+1} = kP_n(L - L) = 0$ so that the population does die out in the next generation. If $P_n > L$, then $L - P_n$ is negative. Therefore, P_{n+1} is the product of a positive term, kP_n, and a negative term, $L - P_n$. So P_{n+1} is itself negative. We interpret this to mean that as before, the population completely dies out in generation $n + 1$. Thus, the rule $P_{n+1} = kP_n(L - P_n)$ satisfies our two assumptions: exponential growth when P_n is small, but disaster when P_n exceeds L. This rule provides us with the iteration rule $x \rightarrow kx(L - x)$. This rule is called the **logistic iteration rule for population growth.**

You should be aware that this is by no means the only iteration rule that satisfies our two assumptions. One of the jobs of the ecologist is to figure out the best mathematical model to use with a given set of assumptions.

THE PARAMETERS

The constants k and L in this rule are **parameters**. They depend on what species we are modeling, the environment, and a lot of other factors. The numbers k and L will be different for populations of rabbits than for populations of elephants. Remember that both k and L will be positive numbers.

As we often do in science to simplify the process of using this model, we use convenient units. In particular, we will always choose units so that $L = 1$. That does not mean our maximum population is one individual. Rather, we may be measuring populations in millions or billions of individuals. Setting L equal to one unit means that our population P_n is always a number between 0 and 1. Thus we think of P_n as measuring the fraction present at generation n of the maximum possible population. For example, $P_n = 1/2$ means that at generation n, the species numbers exactly half the maximum possible population. Similarly, if $P_n = 1/10$, then the population is considerably smaller; it is only one-tenth as large as the maximum population. Thus, the logistic iteration rule becomes simply $x \rightarrow kx(1 - x)$, which is the form of the rule that we will use from now on. This is an example of a nonlinear iteration rule. It is nonlinear because it is not in the form $x \to Ax + B$ like those rules we studied before. If we multiply the rule out, we find that the rule is $x \to kx - kx^2$, which involves the nonlinear term x^2. The presence of this simple squaring term will complicate our lives enormously! We will study quadratic iteration rules like the logistic model extensively in the next few lessons. Before that, however, your job is to gather a little data about the fate of orbits. We suggest that you spend some time with a computer and calculator to perform Investigations 3–6.

NAME(S):

1 ▹ The exponential growth model

a. For the simple exponential growth model $x \rightarrow kx$, for which k-values does the population become extinct?

b. For which k-values does the population increase without bound?

2 ▹ Improving the model

The exponential growth model is much too simple-minded to be of use in actual population prediction. There are a number of situations that this model does not account for. What are some of the things missing from this model that might make it more realistic?

3 ▹ Finding the fixed points

Find the fixed points for the iteration rule $x \rightarrow kx(1 - x)$. Remember that $k > 0$ and that the seed x_0 is a population between 0 and 1. So negative fixed points don't mean anything.

4 ▹ AN EXPERIMENT

COMPUTER OR CALCULATOR INVESTIGATION

Use a computer or calculator to investigate the fate of the orbit of $x_0 = 0.5$ under the iteration rule $x \to kx(1 - x)$ for the given k-values. In each case, iterate long enough so that you feel confident about the fate of the orbit.

a. $k = 0.2$ Fate: ______________________

$k = 0.4$ Fate: ______________________

$k = 0.6$ Fate: ______________________

$k = 0.8$ Fate: ______________________

b. What does the behavior in each of the preceding cases mean in terms of the population of a species?

____

____

____

c. Try some other k-values between 0 and 1. Do you see the same behavior?

____

____

____

d. What can you say about the fate of the orbit of 0 for each of these k-values?

____

____

____

e. What happens when $k = 1$? Or when k is slightly larger than 1, say, 1.1?

____

____

____

5 ▹ Another Experiment

Now consider larger *k*-values.

a. $k = 1.3$ Fate: ______________________

$k = 1.8$ Fate: ______________________

$k = 2$ Fate: ______________________

$k = 2.5$ Fate: ______________________

b. What does the behavior in each of the preceding cases mean in terms of populations?

____

____

____

c. Try some other *k*-values between 1 and 3. Do you see the same behavior?

____

____

____

d. What happens when $k = 3$? Or when k is slightly larger than 3, say, 3.1?

____

____

____

6 ▹ Larger *k*-Values

And even larger *k*-values:

a. $k = 3.2$ Fate: ______________________

$k = 3.4$ Fate: ______________________

$k = 3.5$ Fate: ______________________

$k = 3.554$ Fate: ______________________

b. What does the behavior in each of the preceding cases mean in terms of populations?

____

____

____

c. Try some other k-values between 3 and 3.57. Do you always see the same behavior?

____

____

____

7 ▸ THE ORBIT DIAGRAM

COMPUTER INVESTIGATION

For this activity, you will need access to software that generates orbits or a spreadsheet because you will have to iterate many more times than you can using a calculator.

In order to gain an appreciation of the vast array of behaviors that can occur for the logistic model, you will compute the fates of the orbit of $x_0 = 0.5$ for many different k-values, much as you did in the previous Investigations. This time, however, you will record the results graphically in a picture known as the **orbit diagram**.

a. Select a collection of k-values as described below. (Pick "typical" k-values, like $k = 0.43857$ rather than $k = 1$.) Then compute the fate of the orbit of 0.5 for each of the selected k-values. Iterate enough times so that you can determine if the orbit tends to a fixed point or a cycle. Then record both the chosen k-value and the numerical values along the cycle. For example, if $k = 1.3$, then you should see that the orbit tends to a fixed point at 0.231 . . . , so you would record this k-value and the fixed point 0.231 . . . . If $k = 3.15$, then the orbit tends to a 2-cycle given by 0.533 . . . and 0.784 . . . . (Note that we keep only the first three digits of the cycle.) If $k = 3.5$, then the orbit tends to a 4-cycle at 0.500 . . . , 0.875 . . . , 0.383 . . . , and 0.827. . . . Occasionally, you will be unable to decide if the orbit is actually tending to a cycle or not, in which case just leave that k-value aside for the moment.

Important: Record only the fate of the orbit, not the "transient" behavior. In each case, iterate enough times so that you see the final behavior. If you cannot distinguish the fate, do not worry. Move on to a different k-value and see what happens for this k.

b. When you have gathered your data, plot the results as follows. Plot the chosen k-values on the horizontal axis (between 0 and 3.57). Plot the numbers corresponding to the fates of orbits on the vertical axis. You will plot just one point above $k = 1.3$, the point (1.3, 0.231), which corresponds to a fixed point. Above $k = 3.15$, plot two points, (3.15, 0.533) and (3.15, 0.784), corresponding to the 2-cycle. And there will be four points plotted above $k = 3.5$. The picture thus far looks like the graph at right.

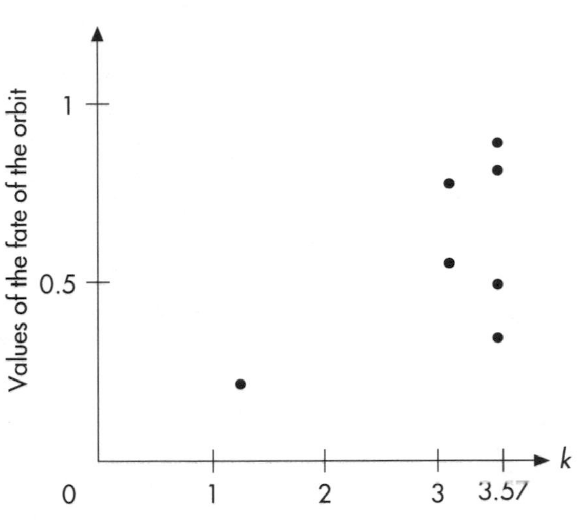

c. To fill in the picture, choose two different k-values in each of the following intervals. Record your choice for k and the fate of the orbit.

$0 < k < 1$

____

____

$1 < k < 2$

____

____

$2 < k < 3$

____

____

$3 < k < 3.44$

$3.46 < k < 3.541$

$3.543 < k < 3.563$

$3.565 < k < 3.568$

d. Now get together with all of your classmates and plot your respective data on the same plot. Does your data agree with that of your classmates? ________________

e. Do you see a pattern here as k tends to 3.57. . . ? In particular, can you find k-values that continue this pattern?

f. Try to paint the picture of the orbit diagram inside the tiny interval of k-values $3.83 < k < 3.85$. Is there any pattern here? Add a magnification to the orbit diagram over these k-values.

1. Carry out a similar analysis of the fate of orbits for the quadratic iteration rule $x \rightarrow x^2 + c$. Here, c is a parameter just like k in the logistic family. You should choose c-values beginning at $c = 0.24$ and decreasing to $c = -1.4$. Investigate the lower regions of this range in more detail. Also, you should always use the seed $x_0 = 0$ rather than 0.5 for this iteration. Do you see any similarities between the orbit diagram for this quadratic family and the logistic family?

TEACHER NOTES

Nonlinear Iteration

Overview

In this lesson, we take a deeper look at nonlinear iteration rules such as the logistic model introduced in Lesson 4. We describe how nonlinear rules may have attracting, neutral, and repelling fixed points. The concept of bifurcation, which will become important in later discussions, is introduced.

Mathematical Prerequisites

Students must have good graphing skills (particularly for quadratic expressions). They must also be familiar with the graphical iteration process, which was introduced in Lesson 3.

Mathematical Connections

These topics relate well to the work with **quadratic functions and graphing** in the second-year algebra curriculum. To find fixed points, students need to solve quadratic equations, so the **quadratic formula** comes into play (especially in Investigation 9). Reference is also made to calculus since determining the slope of a nonlinear iteration rule involves the **slope of the tangent line.**

Technology

Students must have the capability to graph equations accurately, since determining whether a fixed point is attracting or repelling depends on an accurate graph. Ideally, students will use a graphing calculator or other software that performs graphical iteration. The applet "Nonlinear Web" at our Web site **http://math.bu.edu/DYSYS/applets** performs graphical iteration on a number of iteration rules encountered in this book.

SUGGESTED LESSON PLAN

CLASS TIME

One 50-minute period is enough to introduce the concepts and processes, provided you have technology available to illustrate the graphical behavior of seeds around the fixed points. If such technology is not available, you might want to have some of the graphs on transparencies for illustrative purposes. You could use an additional class period for students to work on some of the Investigations in pairs or groups.

PREPARATION

Read through the explanatory material and determine whether you will use technology or transparencies for the visual support needed to present this material. Arrange for students to have access to graphing calculators or computer software that performs graphical iteration.

You might want to review with students the concept of graphical iteration and the idea that the fixed points are the points where the $y = x$ diagonal intersects the graph of the iterative rule.

LESSON DEVELOPMENT

If the previous lesson (Measuring Population Growth) has not been covered, it might be good to begin with Investigations 4 and 5 from that lesson so that students see the fates of certain orbits of the logistic iteration rule before turning to graphical iteration. This might also be a good time to discuss why we limit our investigation to Quadrant 1. (Investigation 3 in Lesson 4 refers to the fact that negative fixed points are meaningless in terms of population.)

Then turn to the question of existence of a fixed point for the logistic iteration rule. Fixed points are found by solving the equation $kx(1 - x) = x$. Although this equation is quadratic, it may be solved without recourse to the quadratic formula since 0 is always a solution.

Note that the other fixed point depends on k. Show this dependence visually by noting how the graph meets the diagonal at different points as k is changed. This in turn emphasizes the relationship between the algebra and the geometry. (It is helpful here to have an overhead display for your graphing calculator or computer to demonstrate how the graph changes as k changes.)

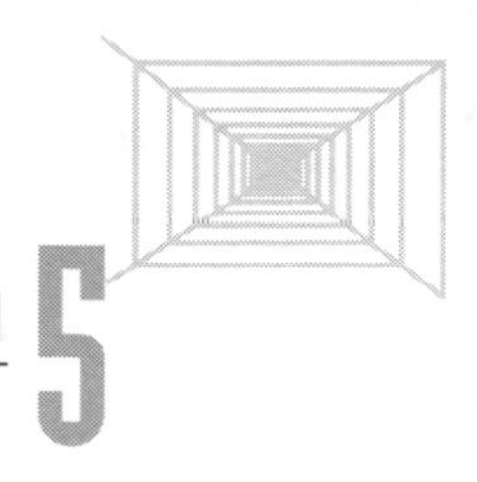

Finally, turn to the question of whether a fixed point is attracting, repelling, or neutral using graphical iteration to determine this. Students will undoubtedly ask, "How do we know what the slope of the graph is? How do we compute it?" This is an ideal opportunity to tell them about calculus. See Lesson Notes below.

The word *bifurcation* first surfaces in this lesson. Point out to students that they will see many more instances of "bifurcations," or changes, in the next few lessons.

Investigations 1, 2, 7, and Further Exploration problems 1 and 2 are fairly straightforward applications of fixed points and graphical iteration for nonlinear iteration. You might want to assign these for homework and leave some of the other Investigations for work during another class period.

LESSON NOTES

The obvious problem that arises in this material is that students have no way of computing the slope of a graph at a point. At this stage, we do two things: We tell our students that this is what calculus does for them. The derivative is the operation that produces the slope of the tangent line (or the slope of the graph) at each point. So, to find out how to do this explicitly, they will just have to take calculus! In the meantime, we show them that graphical iteration or a list of the orbit or a time-series graph determines the nature of the fixed point.

For students who have some background in calculus, you can mention that an attracting fixed point x for the function F occurs when $F(x) = x$ and $|F'(x)| < 1$. Repelling fixed points occur when $|F'(x)| > 1$.

One problem that arises when using a computer to determine the nature of fixed points occurs when the fixed point is neutral. In this case, the convergence to the fixed point is extremely slow. It may take millions of iterations to converge to a given fixed point. On the other hand, the students have the fixed points algebraically, so they can determine their nature by looking at the graph. Also, it is usually (though not always) the case that neutral fixed points herald the appearance of a bifurcation. In practice, we do not worry whether a fixed point is neutral or "weakly attracting" or "weakly repelling." The fact is that the attracting and repelling cases are more important.

Investigations 3 and 4 are designed to reinforce students' skills in finding fixed points and performing graphical iteration, but they take a different approach. When assigning these problems, be sure students understand the directions. Investigations 5 and 6 are designed to reinforce the relationship between the time-series graph and the web diagram.

Investigations 7 and 8 are meant to be fun. They are designed to improve students' comprehension of graphical iteration in an enjoyable manner. These target practice games can be played at our Web site **http://math.bu.edu/DYSYS/applets**.

In fact, finding the cycles in Investigation 8 is almost impossible. There is a good reason for this: Sensitivity to initial conditions—the main ingredient in chaos and the topic of Lesson 7—is the culprit.

Investigation 9 involves an excellent application of the quadratic formula.

In previous lessons, we have seen that the process of graphical iteration provides a geometric tool for understanding the fate of orbits based on linear iteration rules. Here we will see that the same process also works well when the iteration rule is no longer linear. Although we understand completely the fate of orbits for linear iteration, the fate of orbits is much more complicated for nonlinear iterations.

Graphical iteration

Recall that in order to view orbits using graphical iteration, we first superimpose the graphs of our iteration rule and the diagonal $y = x$. The orbit of the seed x_0 is then displayed first by drawing a vertical line from the diagonal to the graph and then by drawing a horizontal line back to the diagonal. The point where this line meets the diagonal sits directly over or under the next point, x_1, on the orbit. To view x_2, we repeat the process: first go vertically to the graph, then go horizontally to the diagonal.

For example, suppose the iteration rule is $x \to x^2$. This iteration rule has two fixed points, at 0 and at 1. The orbit of the seed 0.9 is

$$0.9 \to 0.81 \to 0.6561 \to 0.4305 \ldots \to 0.1853 \ldots \to 0.034 \to \cdots$$

This orbit tends away from the fixed point at 1 and toward the fixed point at 0. We see this clearly in the staircase leading down to 0 in the graphical iteration that follows.

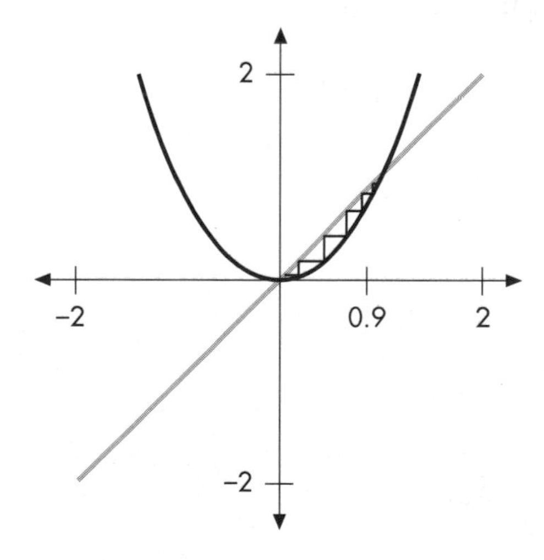

On the other hand, the orbit of 1.01 tends away from 1 and toward infinity while the orbit of –0.9 tends to 0, as shown here:

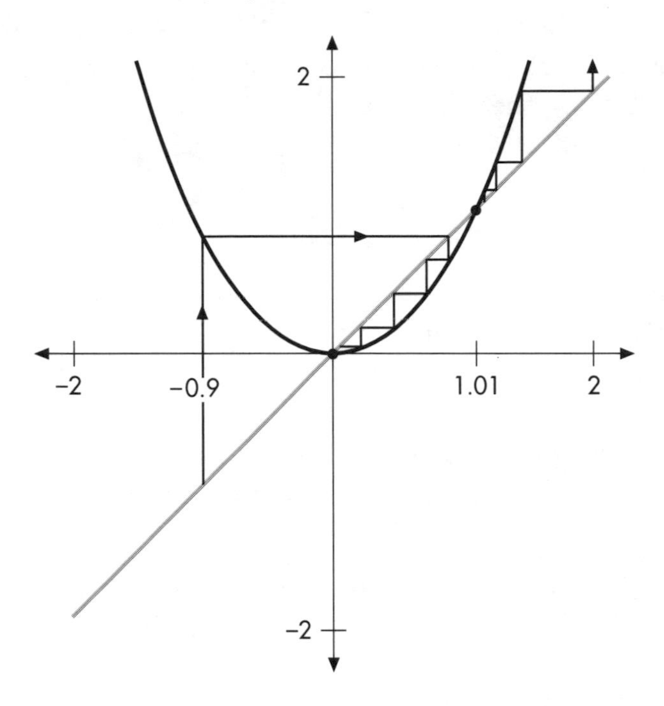

TYPES OF FIXED POINTS

Here is the graph of a nonlinear iteration rule. This graph crosses the diagonal in two places, so we have two fixed points marked *A* and *B* for this iteration. We have sketched the orbits of several seeds for this iteration. Note that the orbits of x_0 and y_0 move away from the fixed point at *A*, while the orbits of y_0 and z_0 move toward the fixed point at *B*. We therefore call the fixed point *A* repelling and the fixed point *B* attracting.

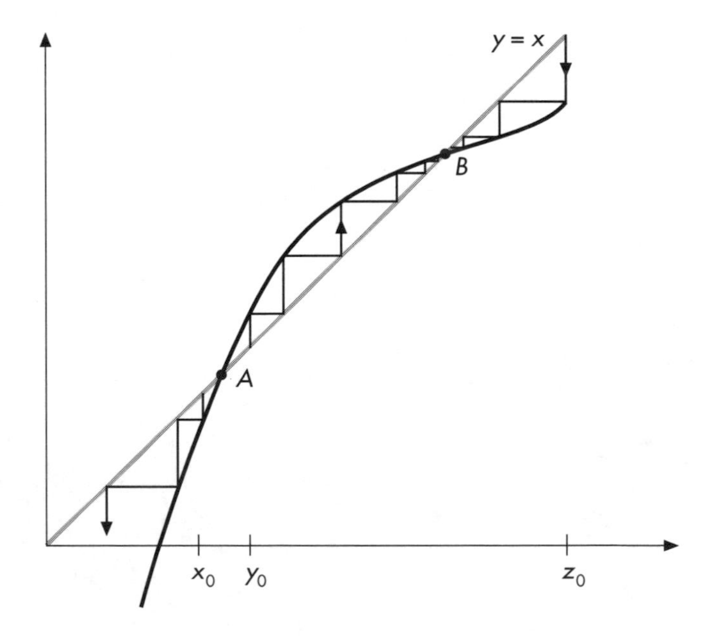

This is a little different from the situation for linear iteration rules. In the linear case, *all* orbits tended to an attracting fixed point. In the nonlinear case, all we can hope for is that seeds that are sufficiently close to an attracting fixed point have orbits that tend to the fixed point. Similarly, seeds located near a repelling fixed point tend to move away from it.

The logistic function

Now let's return to the logistic iteration rule. The graph of $y = kx(1 - x)$ assumes several different shapes depending on how large k is. Here are several graphs. Recall that we are concerned only with values of x between 0 and 1.

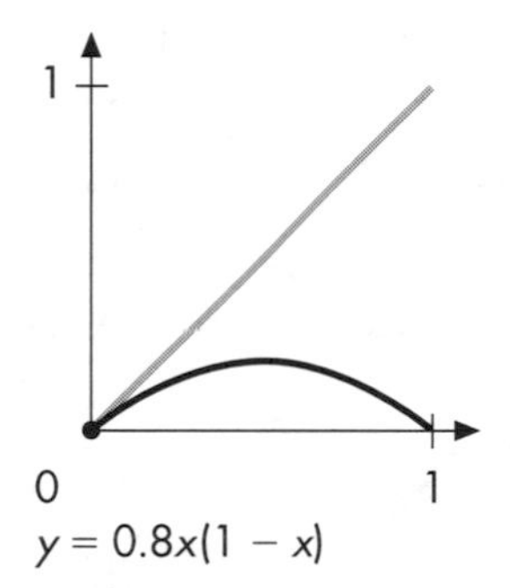

$y = 0.8x(1 - x)$

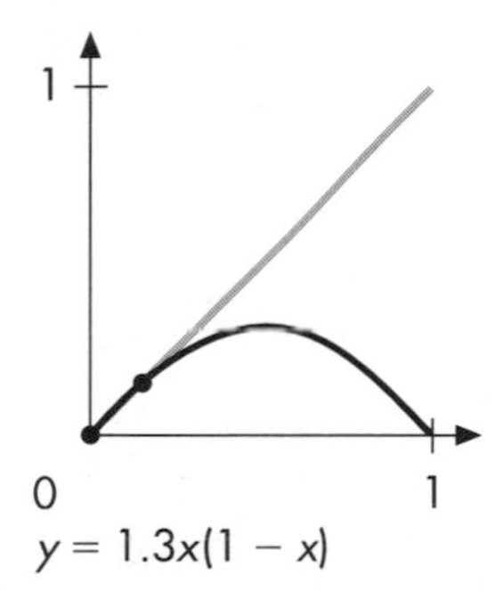

$y = 1.3x(1 - x)$

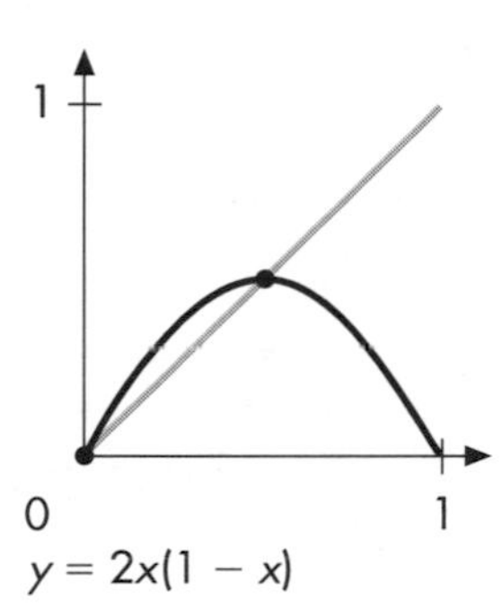

$y = 2x(1 - x)$

When $k = 0.8$, we see that the graph crosses the diagonal just one time in the interval $0 \leq x \leq 1$, but when $k = 1.3$ or $k = 2$, the graph crosses the diagonal twice. So we have only one fixed point for $k = 0.8$ in the interval $0 \leq x \leq 1$, but there are two fixed points for $k = 1.3$ or $k = 2$. At what value of k does this change take place? We can answer this by solving the equation for the fixed points. We need to know when $kx(1 - x) = x$.

One solution of this equation is $x = 0$. We always have a fixed point at 0. To find the other fixed points, we first divide both sides by x. We can do this since $x \neq 0$. We find

$$k(1 - x) = 1 \quad \text{or} \quad -kx = 1 - k$$

Solving for x yields

$$x = \frac{1 - k}{-k} \quad \text{or} \quad x = \frac{k - 1}{k}$$

Now recall again that k and x are always positive since we are dealing with populations. The denominator in this expression is always positive, so to find a second fixed point that is positive, we need the numerator to be positive too.

Since the numerator is $k - 1$, we must have $k > 1$. In that case, our second fixed point is the positive number

$$\frac{k-1}{k}$$

Therefore, the logistic iteration has only one (non-negative) fixed point when $0 < k < 1$ and two fixed points when $k > 1$.

Not only does the number of fixed points for the logistic iteration rule change when k increases through 1, but the type of the fixed point changes as well. When $k < 1$, 0 is an attracting fixed point for this rule, as we see from graphical iteration.

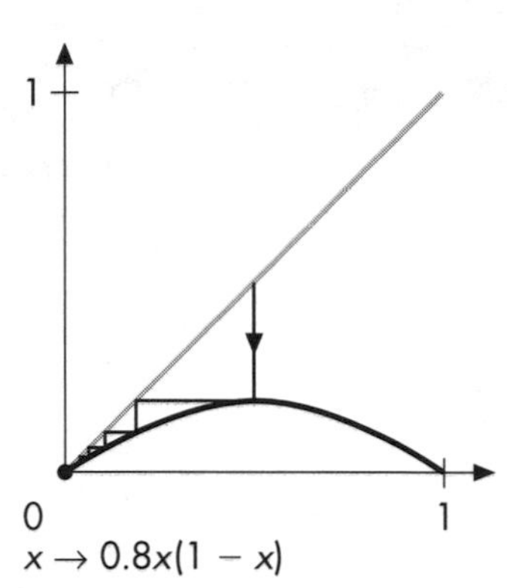

$x \to 0.8x(1 - x)$

The orbit of any x between 0 and 1 must always decrease toward 0. When $k > 1$, the fixed point at 0 becomes repelling, whereas the new fixed point (at least for small k-values) becomes attracting, as we see at right.

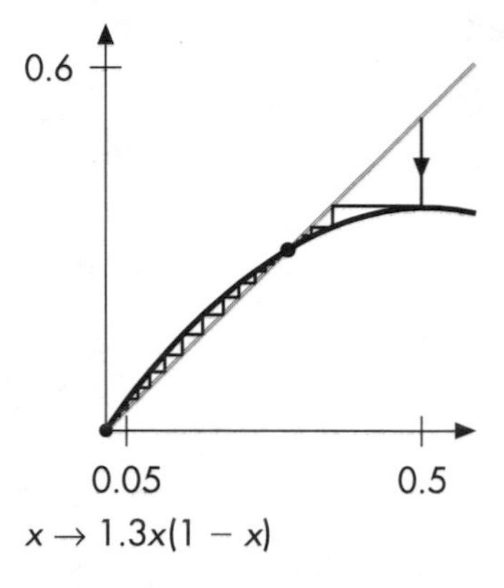

$x \to 1.3x(1 - x)$

Bifurcation

The word **bifurcation** means a change, a splitting apart. We say that the logistic iteration rule undergoes a bifurcation when the parameter $k = 1$ since a new fixed point "splits apart" from the fixed point at 0 as k increases. The study of bifurcations is one of the most important areas of dynamical systems. How and when physical, chemical, and biological systems undergo sudden changes of behavior is also the subject of bifurcation theory.

DETERMINING THE TYPE OF FIXED POINTS

To determine whether a fixed point is attracting or repelling, we must carefully examine the position of the graph relative to the diagonal. Recall what happened for a linear iteration rule $x \rightarrow Ax + B$. In this case, the graph of the iteration rule was a straight line with slope A. If $|A| > 1$, then we saw that the fixed point was repelling. If $|A| < 1$, then the fixed point was attracting. If $|A| = 1$, then the fixed point was neutral. In summary, it was the slope of the graph of the iteration rule that controlled the behavior of orbits near the fixed point.

The same is true for nonlinear iteration rules. If the graph crosses the diagonal with slope greater than 1, then the fixed point must be repelling. But if the slope of the graph at the fixed point is between 0 and 1, then the fixed point is attracting.

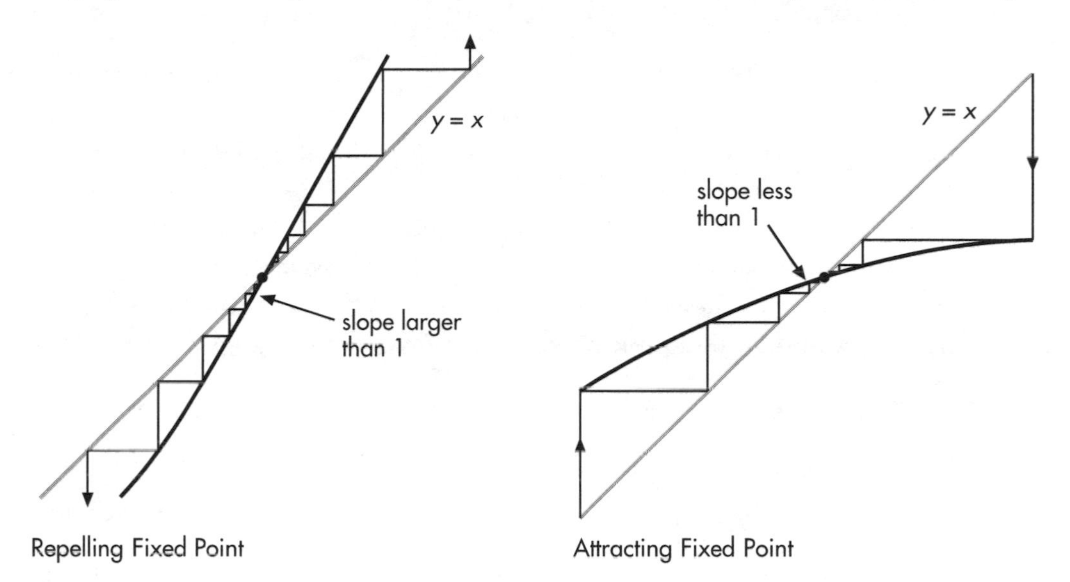

Repelling Fixed Point

Attracting Fixed Point

The slopes at fixed points can also be negative. If the slope is "not too negative," the fixed point will be attracting, but if the slope is very steep and negative, the fixed point will be repelling.

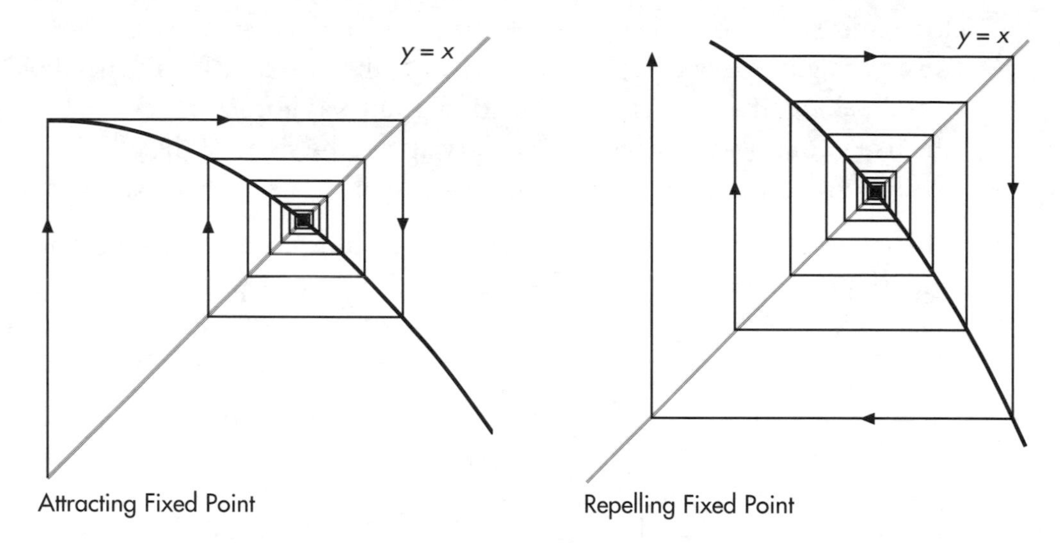

To determine where the fixed point changes from attracting to repelling, we zoom in on the fixed point. Here are three successive zooms on the graph of $y = 2.8x(1 - x)$. As we zoom in, the graph of the iteration rule itself becomes straighter and straighter, looking more and more like a straight line.

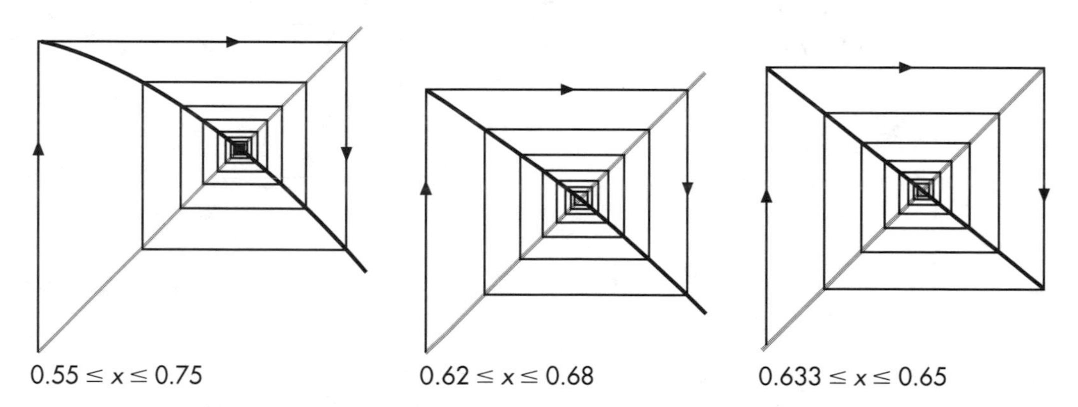

Now, from what we observed in Lesson 4, the fixed point of a linear iteration rule is attracting if the slope is between -1 and 1, and repelling if the slope is less than -1. So the same should be true for nonlinear iteration rules.

To summarize, a fixed point of a nonlinear iteration rule is attracting if the slope of the graph at the fixed point is between -1 and $+1$. It is repelling if the slope is either larger than $+1$ or less than -1. Of course, the big question is: How do you compute the slope of the tangent line to the graph of a nonlinear expression? Luckily, this is exactly the question that calculus answers. When you study differential calculus, you will see that differentiation (one of the most important operations in calculus) gives precisely the slope we desire.

NEUTRAL FIXED POINTS

As in the case of linear iterations, there are some special cases when the graph has a fixed point at which the slope is either $+1$ or -1. In these cases, the fixed point may be attracting, repelling, or neither, in which case we have a **neutral fixed point**. For example, the graph of $y = x^2 + x$ has a fixed point at $x = 0$. From the right, this fixed point looks repelling, but from the left it appears to be attracting. Therefore, 0 is neither attracting nor repelling: It is a neutral fixed point.

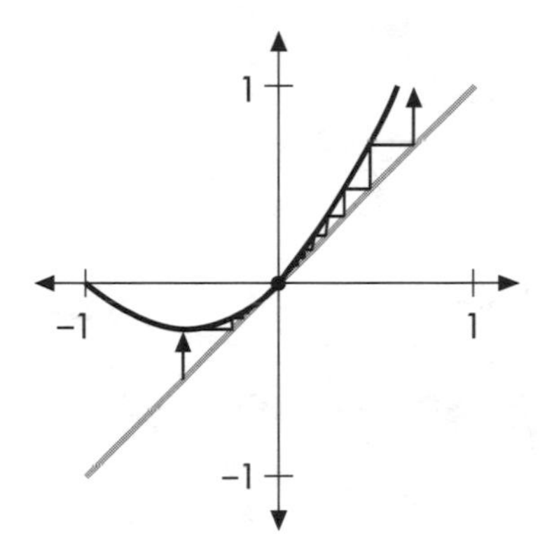

1 ▹ GRAPHICAL ITERATION

You do not need a formula in order to carry out graphical iteration. For example, here is the graph of a nonlinear iteration rule. Using the diagonal line and graphical iteration, sketch several orbits for this iteration, then answer the questions.

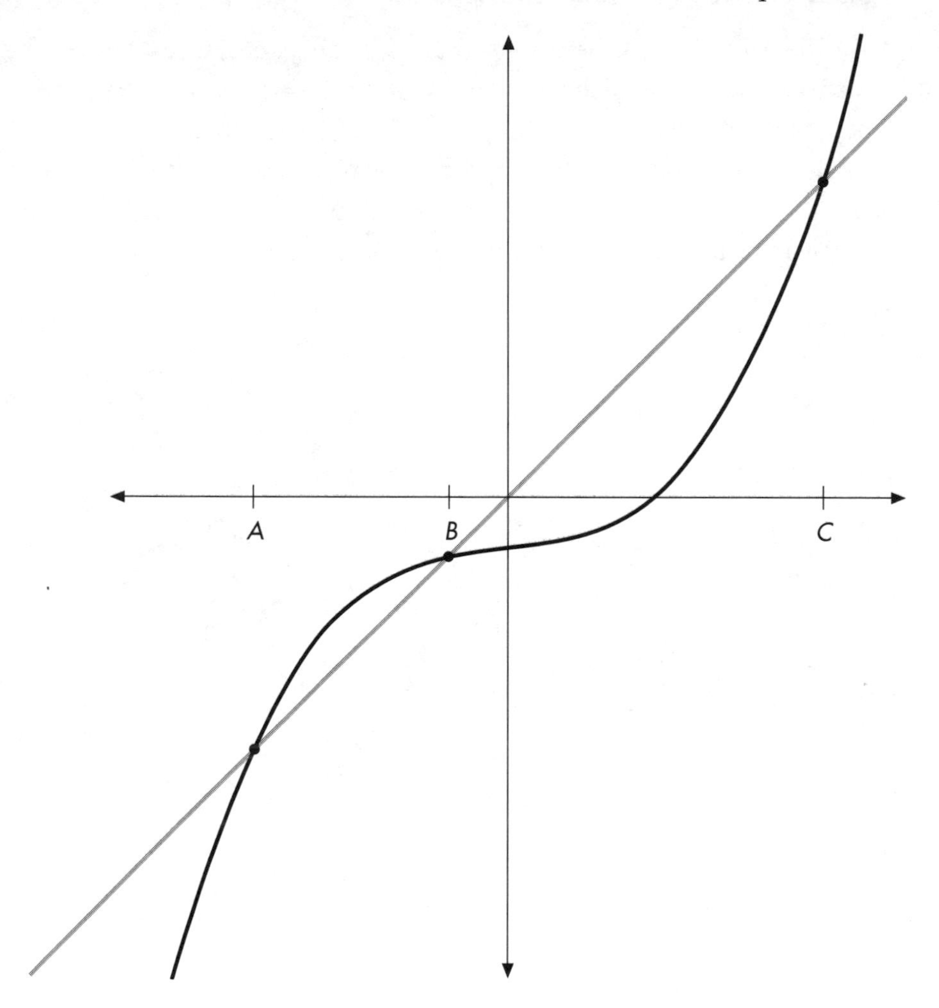

a. Note that this function has three fixed points, at *A*, *B*, and *C*. What is the fate of orbits that begin with seed $x > C$?

____

____

b. What happens if $B < x < C$?

____

____

c. What happens if $A < x < B$?

d. What happens if $x < A$?

e. Which of the three fixed points are repelling? Which are attracting?

2 ▹ THE LOGISTIC ITERATION RULE

For each of the following logistic iteration rules, first find all fixed points. Then use graphical iteration to determine if the fixed points are attracting or repelling.

a. $x \rightarrow 2x(1 - x)$

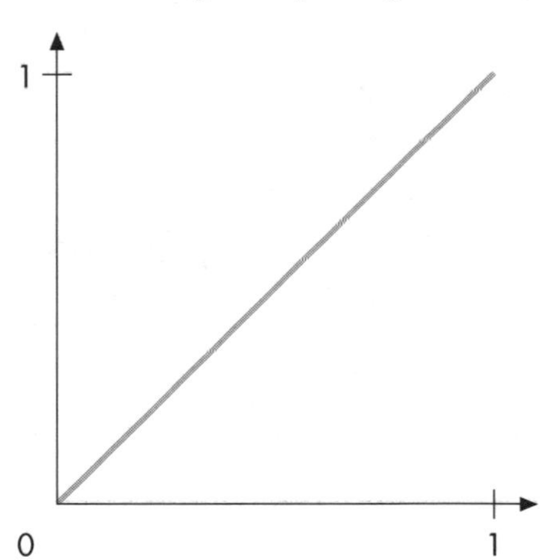

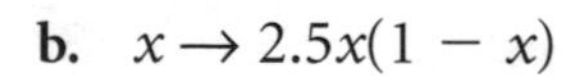

b. $x \rightarrow 2.5x(1 - x)$

____

____

____

____

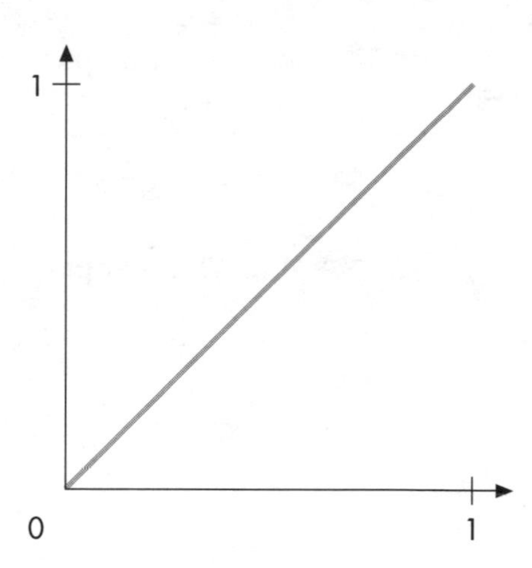

c. $x \rightarrow 3.2x(1 - x)$

____

____

____

____

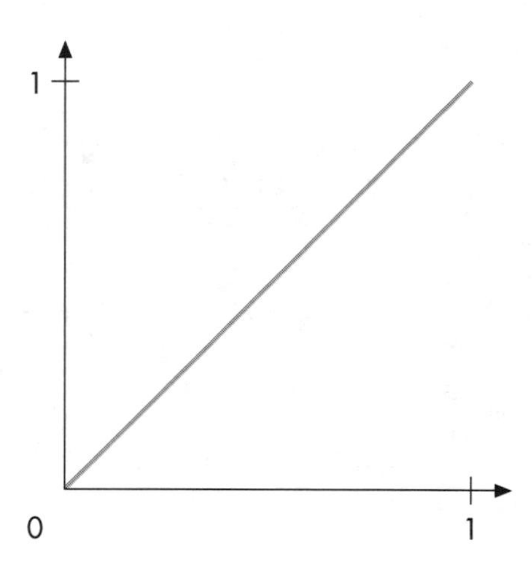

d. There is a difference between the behavior of orbits in parts a and b and the behavior in part c. What change has occurred?

____

____

3 ▹ OTHER ITERATION RULES

Each of the following iteration rules has a fixed point at 0. First, sketch the graph of the given iteration rule. Then, using graphical iteration, determine whether this fixed point is attracting, repelling, or neutral.

a. $x \rightarrow 2x - x^2$ ________________

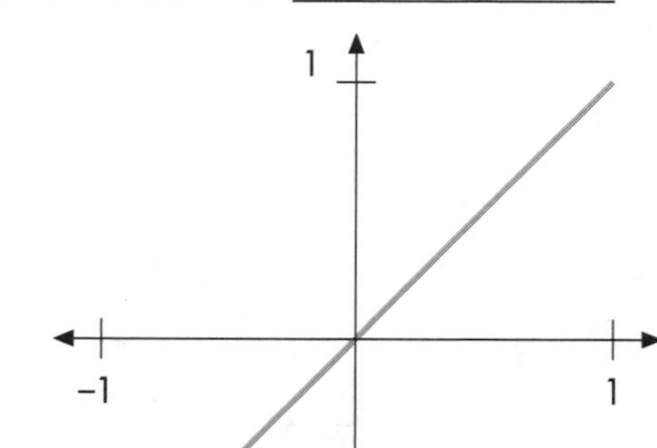

b. $x \rightarrow x^2 + \frac{x}{2}$ ________________

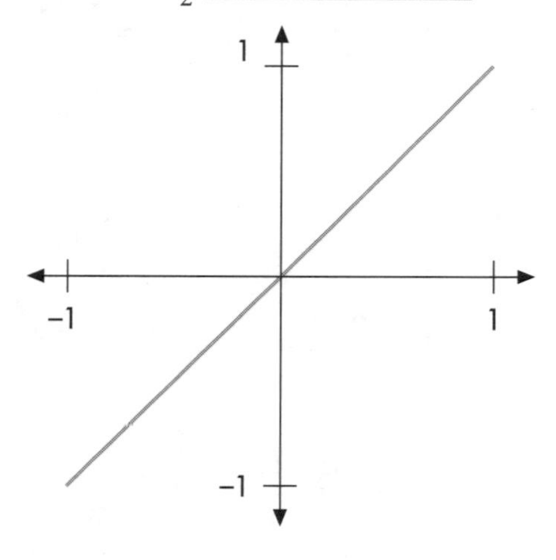

c. $x \rightarrow x^2 - \frac{x}{2}$ ________________

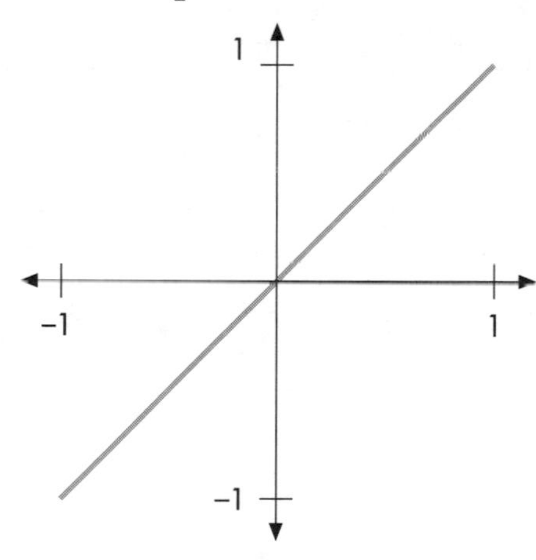

d. $x \rightarrow -2x - x^2$ ________________

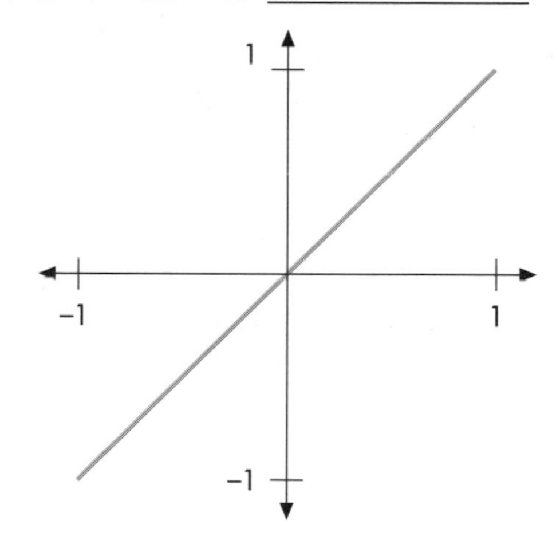

4 ▹ DETERMINING THE FATE OF ORBITS GRAPHICALLY

Here is the graph of the iteration rule $x \to \sqrt{x}$. Using graphical iteration, determine the fate of all orbits. Summarize your findings.

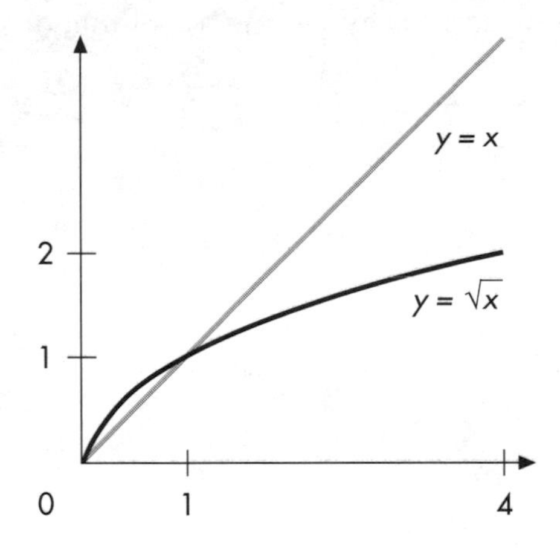

Summary: ____

____

____

____

5 ▹ GRAPHICAL ITERATION AND TIME-SERIES GRAPHS

Here are the graphical iterations corresponding to several different seeds for the iteration rule $x \rightarrow 3.3x(1 - x)$. In each case, sketch the (approximate) corresponding time-series graph for this orbit.

a.

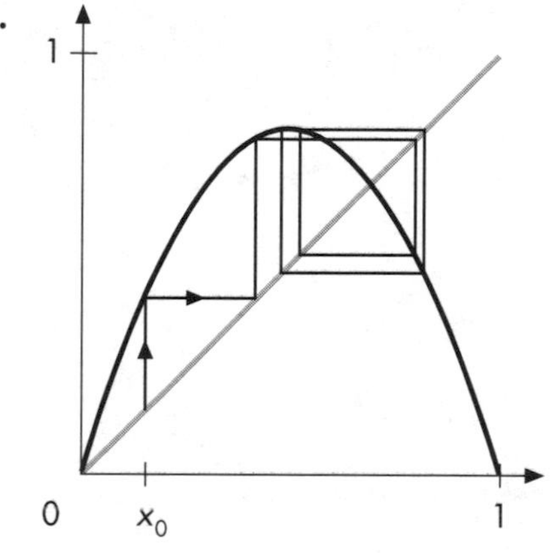

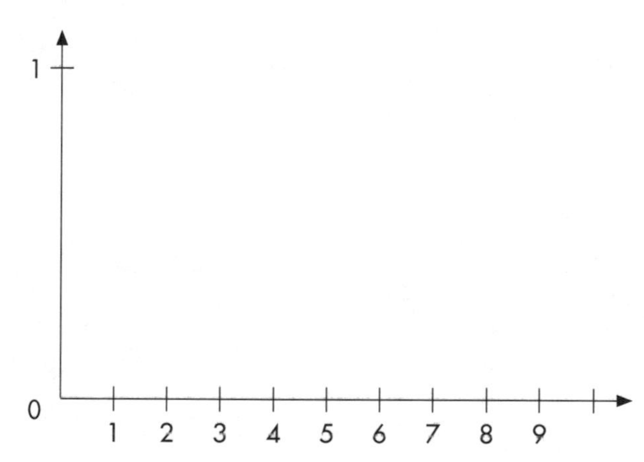

b.

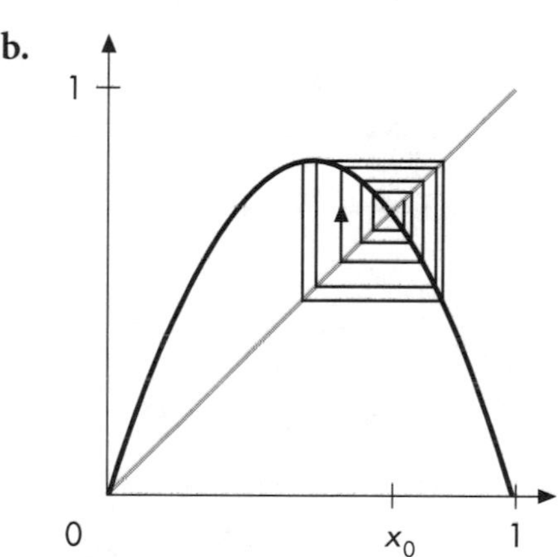

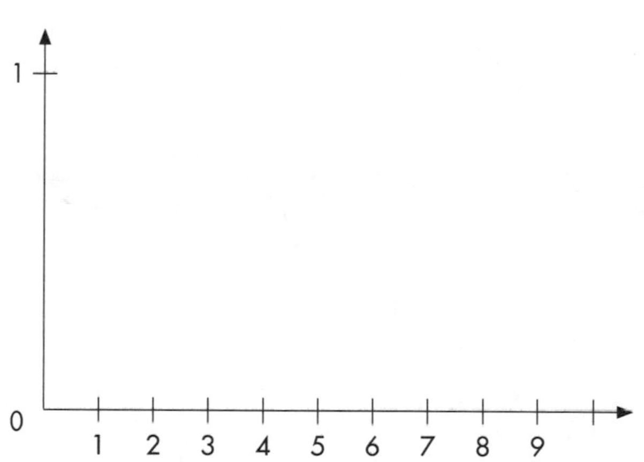

c.

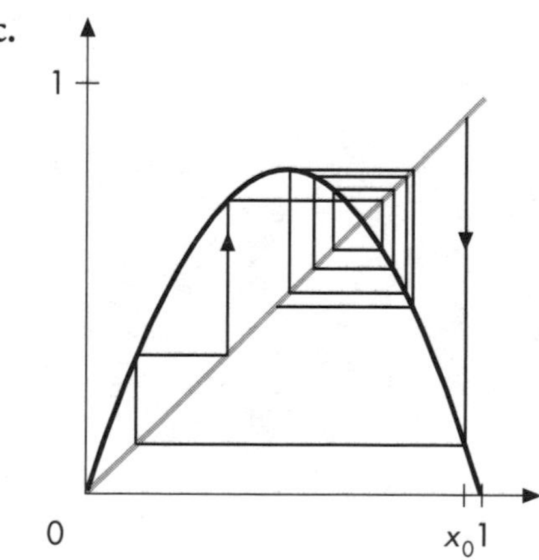

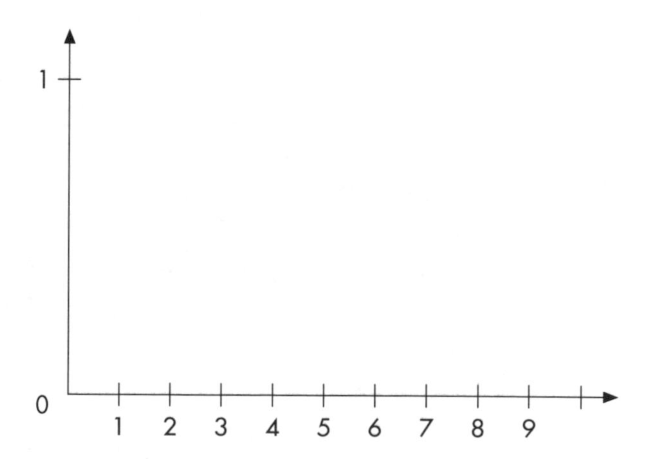

6 ▹ MATCHING GRAPHS

Here are six time-series graphs and six graphical iterations. Match the correct graphs.

A. ____________

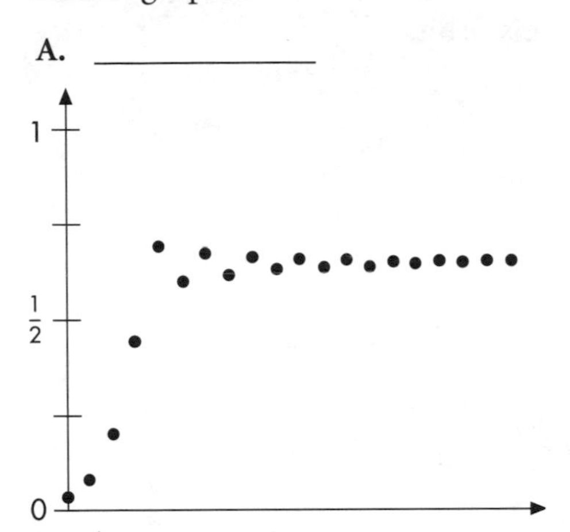

B. ____________

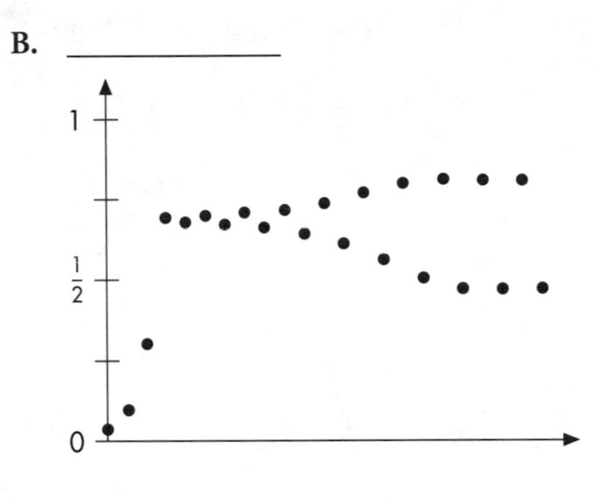

C. ____________

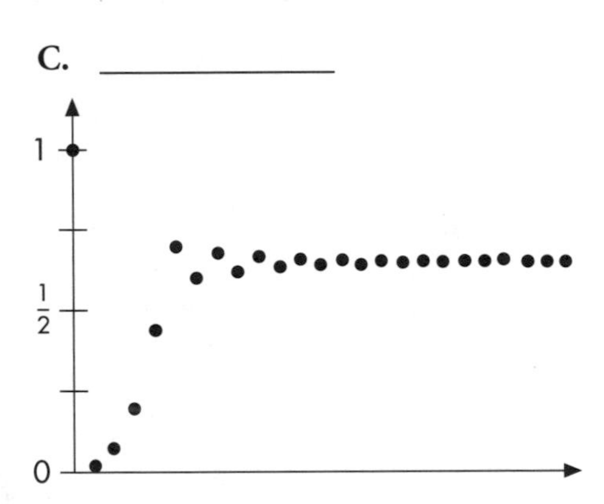

D. ____________

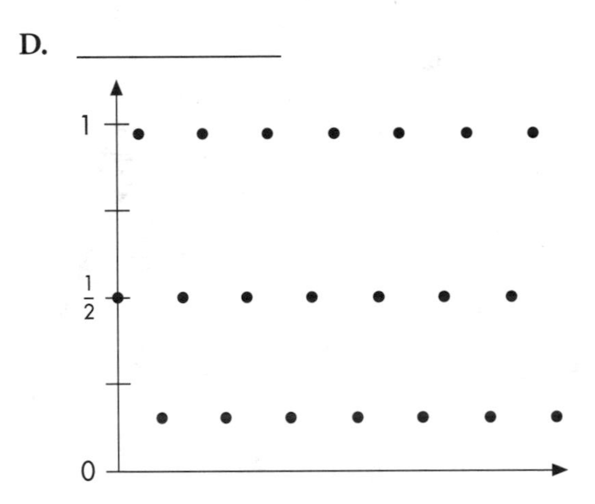

E. ____________

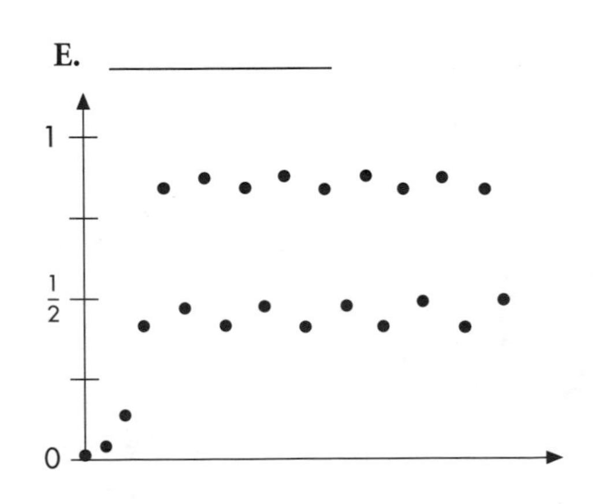

F. ____________

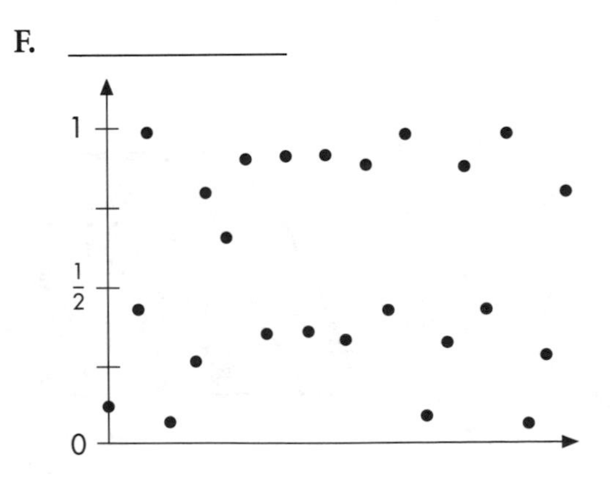

I.

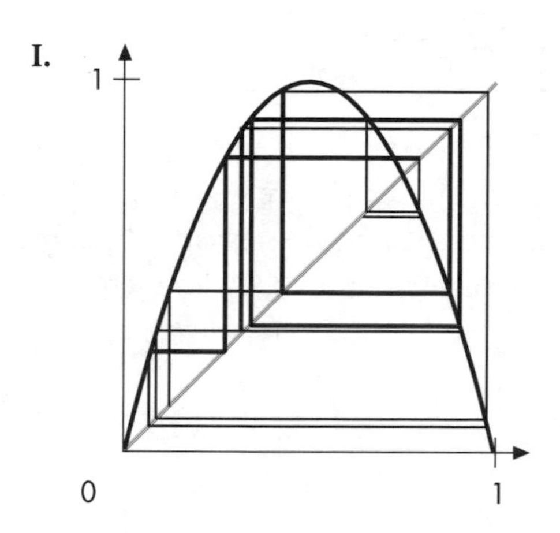

II.

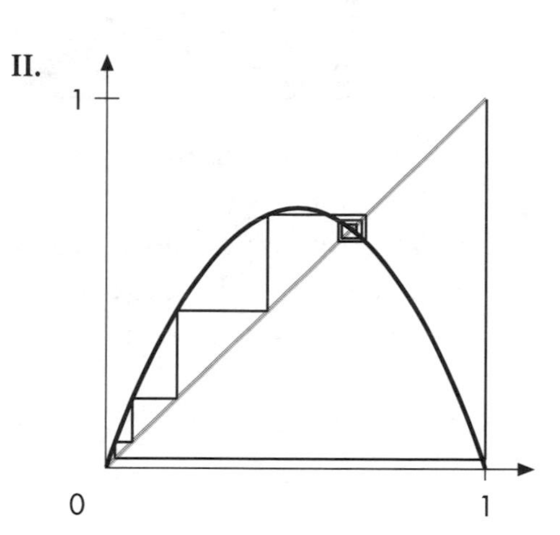

III.

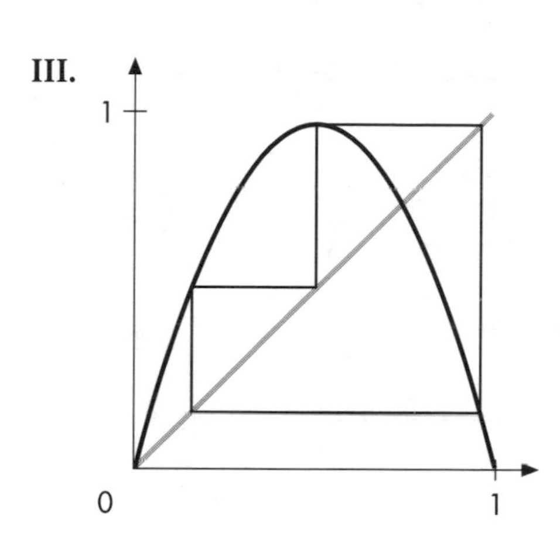

IV.

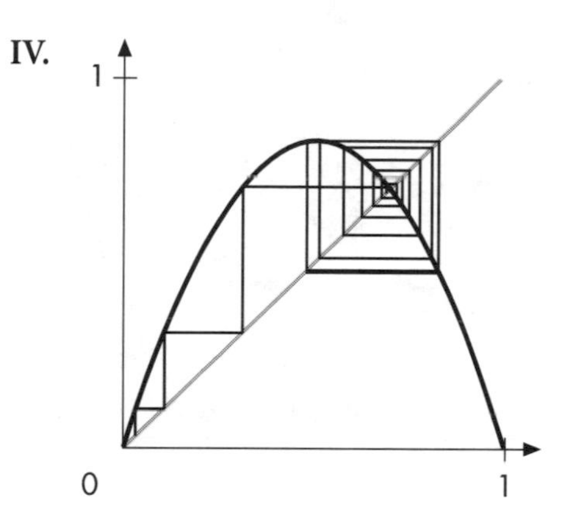

V.

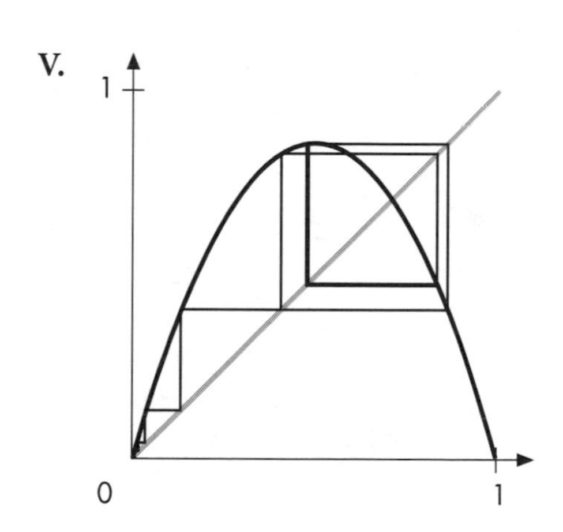

VI.

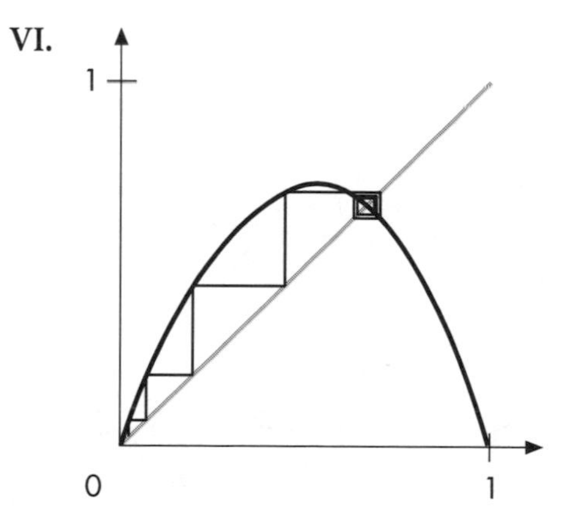

7 ▹ TARGET PRACTICE

Below is the graph of $y = 4x(1 - x)$ for $0 \leq x \leq 1$. A is the interval $0 \leq x \leq \frac{1}{2}$ and B is the interval $\frac{1}{2} \leq x \leq 1$. We give you a sequence of A's and B's and your job is to find an orbit that visits the intervals A and B in exactly that order. The sequence is called the **itinerary** of the seed. For example, for the sequence ABA, the seed x_0 has the correct itinerary.

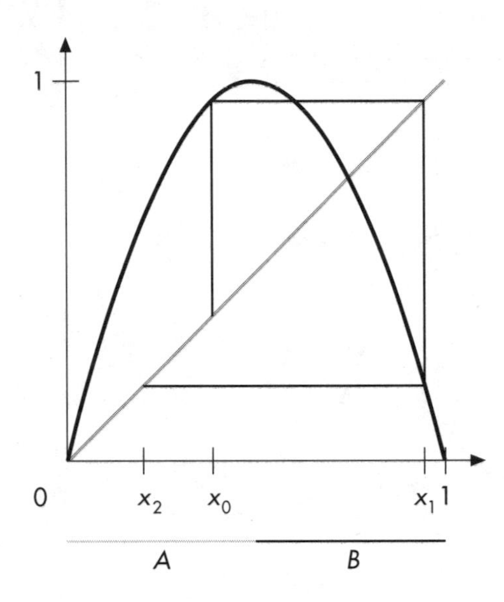

Sketch an orbit corresponding to each of the following itineraries using graphical iteration.

a. AABB

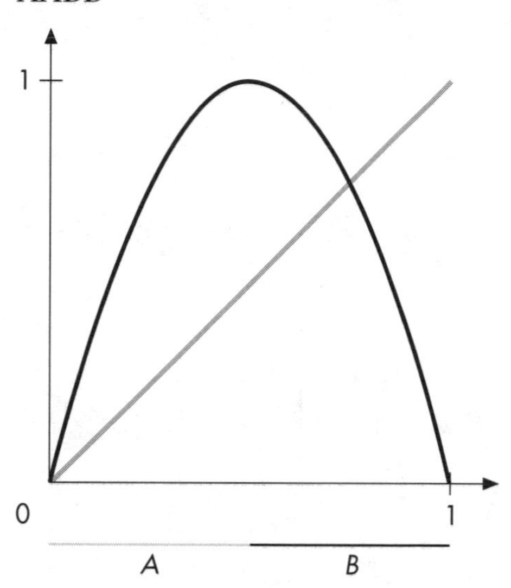

b. ABBAB

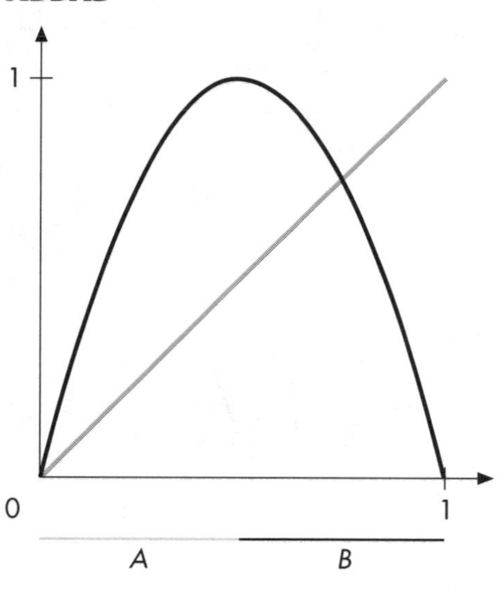

c. BBBBA

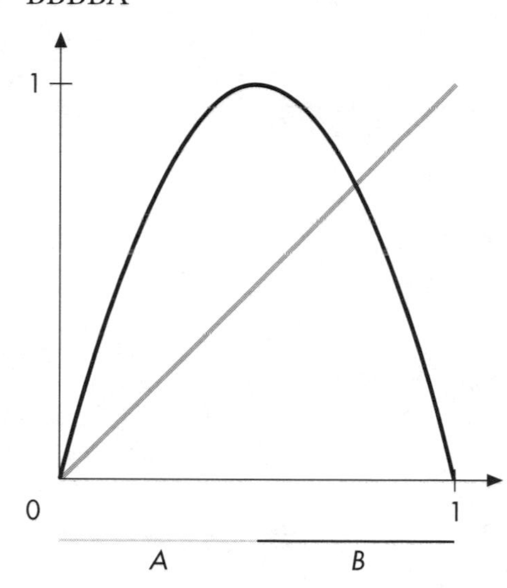

TECHNOLOGY TIP

You may use the following applet to help you find these orbits:

http://math.bu.edu/DYSYS/applets/target.html

8 ▹ CYCLE PRACTICE

Each of the following itineraries corresponds to a cycle for the iteration rule $x \rightarrow 4x(1 - x)$. Use graphical iteration to sketch the orbit of a cycle that has the given itinerary.

a. BBBB . . .

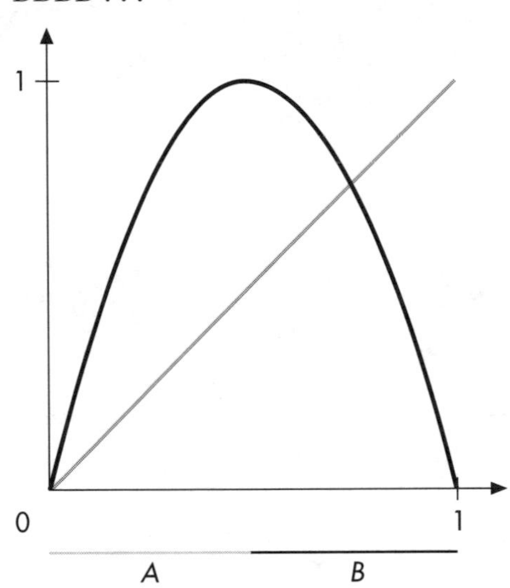

b. ABABAB . . .

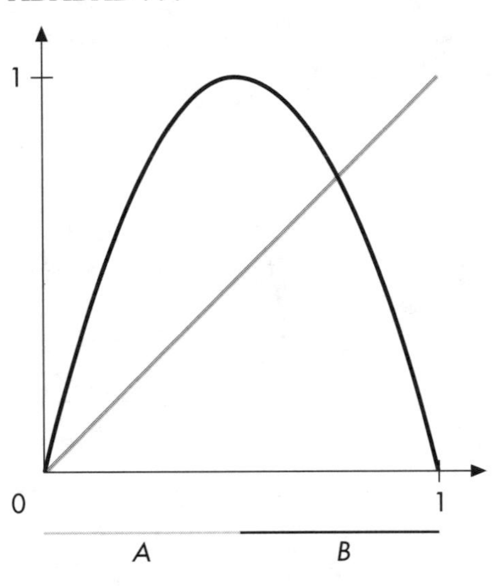

c. AABAAB . . .

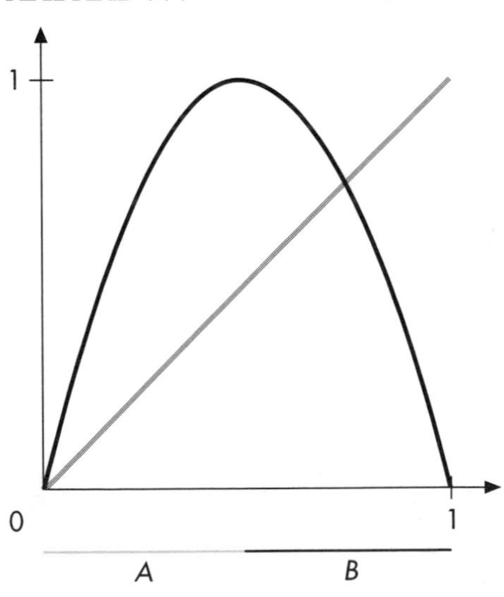

d. ABBABB . . .

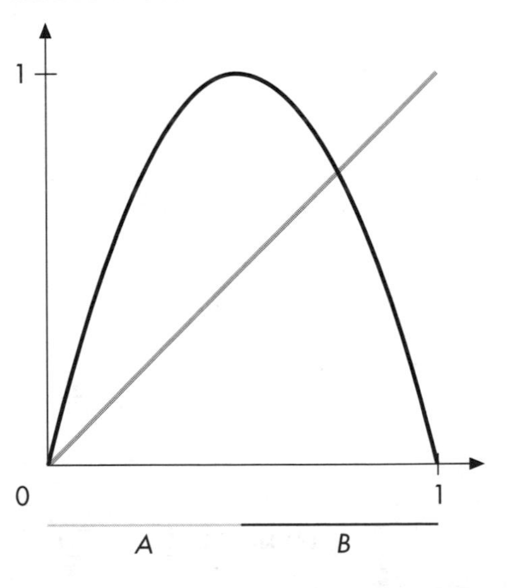

9 ▹ A QUADRATIC ITERATION RULE

Consider the iteration rule $x \rightarrow x^2 + c$ for various values of c. The quantity c is a parameter here. First use algebra to find all c-values for which this iteration rule has

a. No fixed points:

____

____

b. Only one fixed point:

____

____

c. Two fixed points:

____

____

d. At which c-value does this iteration rule experience a bifurcation?

____

____

e. Next use a calculator or computer to determine the c-values (at least approximately) for which the fixed points in part c are attracting and/or repelling. ____________

f. Sketch the graphical iteration corresponding to several of the cases you find in part e.

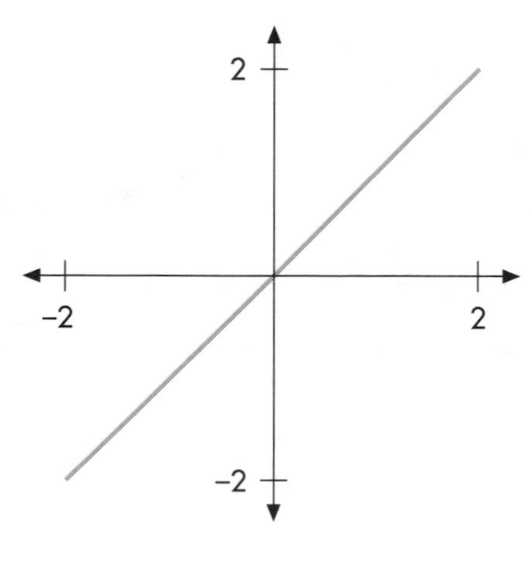

$c =$ ____________

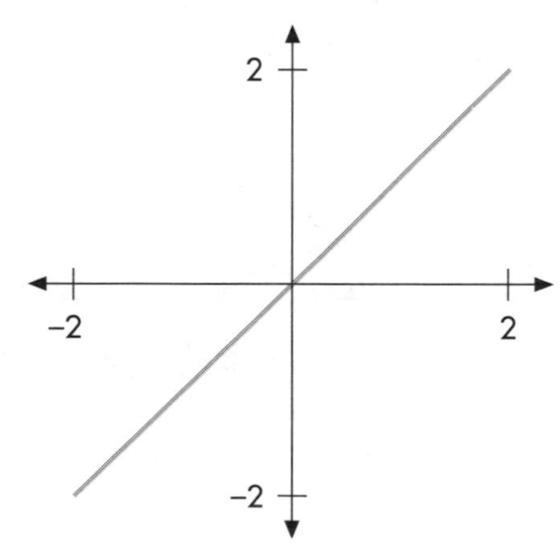

$c =$ ____________

1. For each of the following iteration rules, first sketch the graph of the rule using graphical iteration, then determine the fate of all orbits. Find any fixed points and determine if they are attracting, repelling, or neutral.

   a. $x \to x^3$

   b. $x \to x^2 + 1$

   c. $x \to \frac{1}{x}$

2. Using graphical iteration, determine the fate of all orbits of the iteration rule whose graph is shown below.

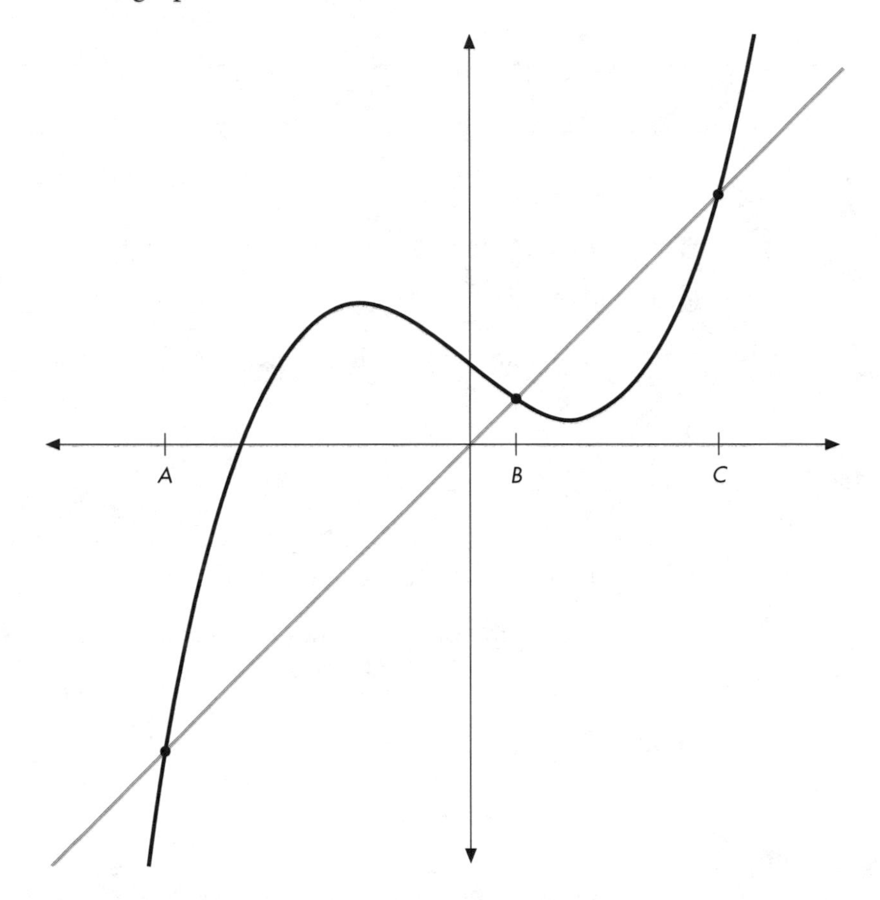

   Which of the three fixed points are repelling? Which are attracting?

3. Use graphical iteration to discuss the fate of all orbits of the iteration rule $x \to \cos x$.

4. Use graphical iteration to discuss the fate of all orbits of the iteration rule $x \to \sin x$.

5. Use graphical iteration to discuss the fate of all orbits of this iteration rule:

$$x \to \begin{cases} \frac{1}{2}x & \text{if} \quad x \geq 0 \\ \frac{1}{2}x + 1 & \text{if} \quad x < 0 \end{cases}$$

Teacher Notes
Chaos

Overview

In this lesson, we get our first glimpse of chaos. We use the familiar tools of time-series graphs and web diagrams to view chaos in the logistic iteration rule $x \rightarrow 4x(1 - x)$. Then we introduce a third way to visualize orbits, the histogram. During the analysis we encounter one more type of fixed point—an eventually fixed point.

Mathematical Prerequisites

It is helpful, though not necessary, if students are familiar with histograms at the outset.

Mathematical Connections

Using the **histogram** to display the data comprising orbits provides a connection to statistics. The addition of the histogram to our iteration tools further reinforces the NCTM *Standards'* emphasis on **multiple representations** of mathematical concepts and the connection between numerical and graphical representations.

Technology

Students need technology in order to draw histograms that consist of more than 30 or 40 data points. Graphing calculators with a table feature or spreadsheets can be used to paint histograms of several hundred points on the orbit. There is an applet available at our Web site to draw histograms consisting of thousands of data points.

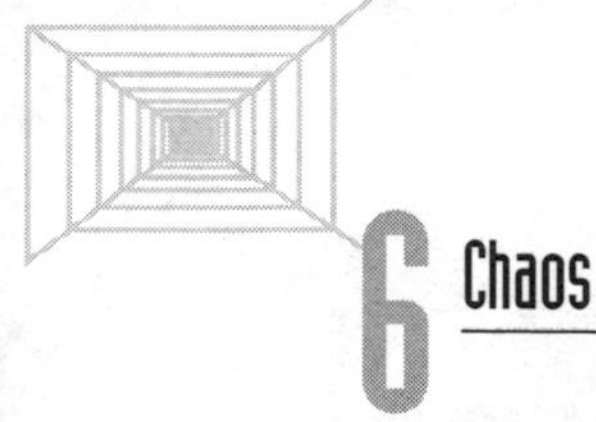

Suggested Lesson Plan

Class Time

One 50-minute class period is enough provided the sample orbits can be traced on graphing calculators or computer software, or transparencies can be made of the various orbits.

Preparation

You may use the transparency masters that follow to have students begin to fill in histograms by hand. It is also advisable to arrange for students to use technology once they have attempted to compute the orbits manually.

Lesson Development

The main goal of this lesson is to have students comprehend the relationship between the three different pictures of orbits that we now have: time-series graphs, web diagrams, and histograms. Students should see and comprehend the difference between a "chaotic" histogram and one in which the orbit tends to a fixed point or cycle. The material on determining the *c*-value from the shape of the histogram is optional.

Investigations 1 and 2 require students to begin to draw a histogram by hand. Investigation 3 is an ideal way to show the connection between a time-series graph and a histogram; Investigation 5 connects histograms with web diagrams.

If time permits, have students work on Investigation 1 in class to be sure they understand this histogram technique; then assign Investigations 2–5 for homework. (Students will need access to a graphing calculator or computer technology to do the homework.)

Lesson Notes

The concept of transient behavior is introduced briefly, but students need to know what it refers to. In more complex systems, like the ones we are looking at in this lesson and beyond, it often takes a while for a seed to "settle in" to its final fate. Often, the initial "noise" or transient behavior interferes with being able to recognize the final orbit. So mathematicians often "throw away" the first 50 to 100 iterations in order to see the true fate of the later orbit.

If you plan to cover the latter sections of this lesson, be sure to read carefully the statements regarding how to read off the value of *c* from the histogram. You need to follow the forward part of the orbit of 0.5 in this process, not including the seed 0.5. Then the peaks of the histogram occur more or less in the order of this orbit. Since this is a statistical process, your results may vary somewhat.

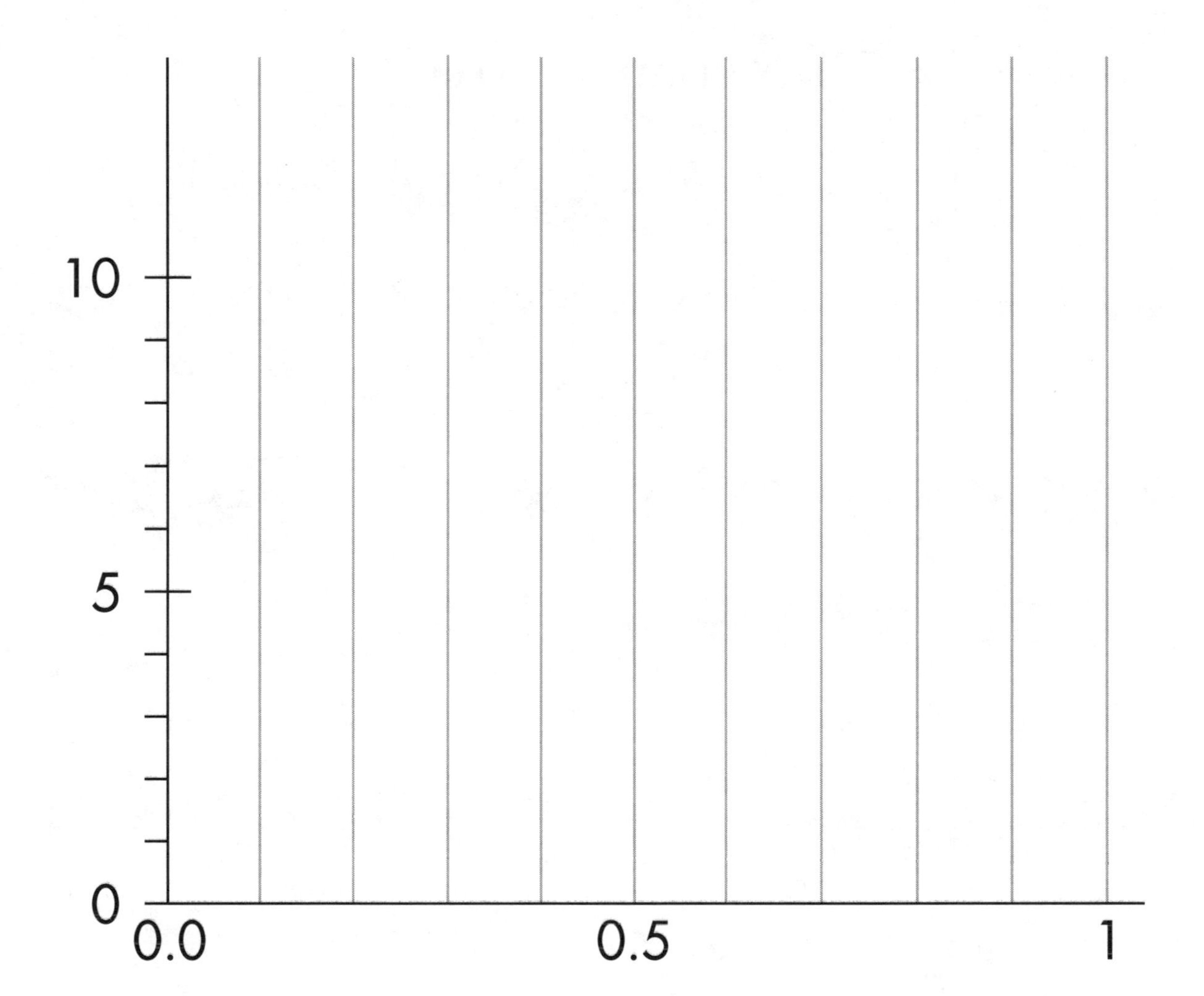
10
5
0
0.0
0.5
1

Histogram for $4x(1 - x)$

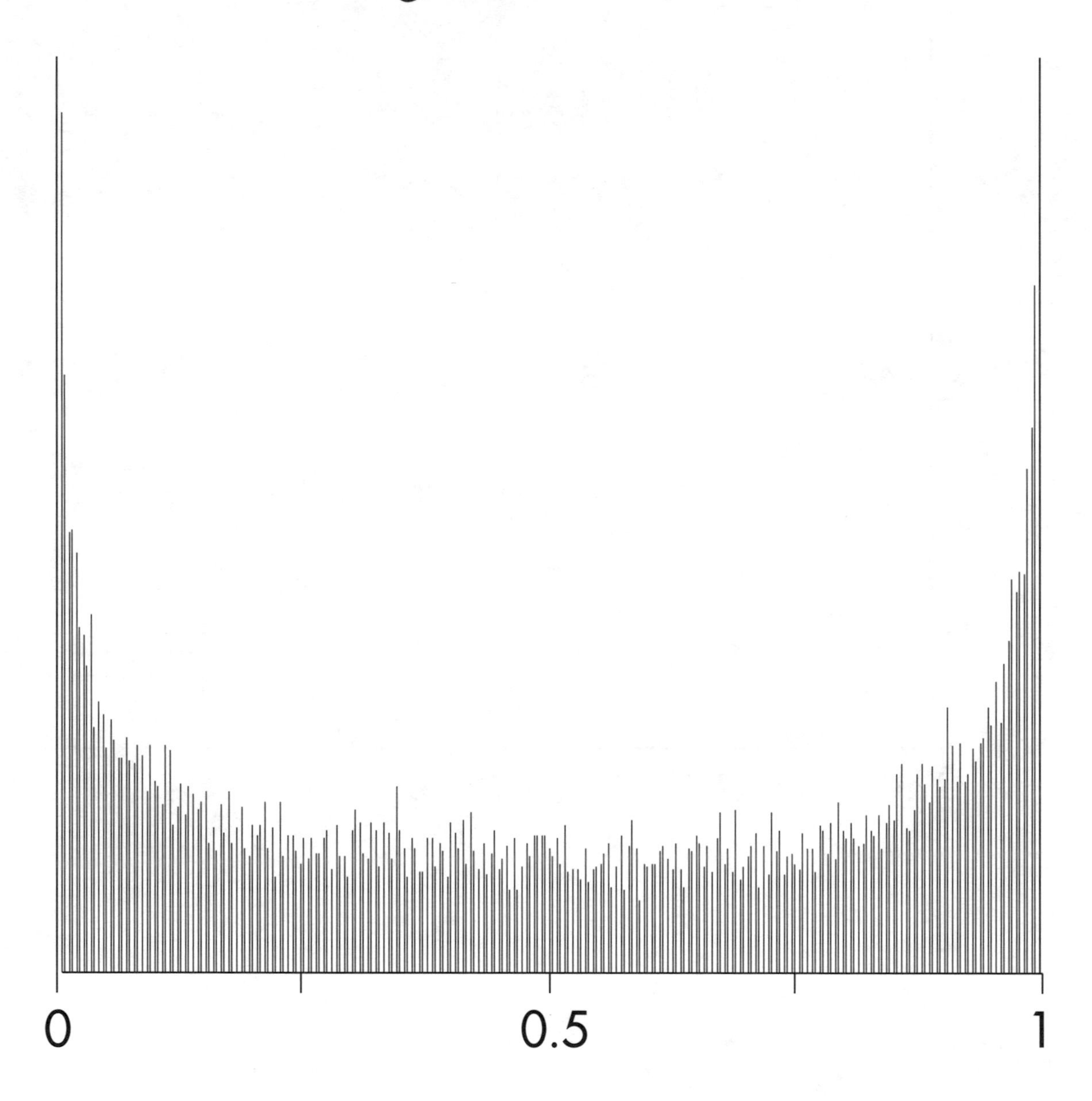

In the previous lesson (Nonlinear Iteration), we used the computer to investigate numerically the fate of orbits of the logistic iteration rule $x \to kx(1-x)$ for various values of the parameter k. We found fixed points for this rule and examined orbits via graphical iteration. The picture that emerged in that lesson was, for the most part, much like what we saw for linear iterations. However, as we will see in this lesson, nonlinear iteration rules can have much more complicated behavior. The culprit is chaos, the topic of this lesson.

A SPECIAL CASE

We first consider a special case of the logistic iteration rule where $k = 4$, so $x \to 4x(1-x)$.

To find the fixed points of this iteration, we must solve the equation $4x(1-x) = x$. One solution is $x = 0$. Factoring out x, we then find the equation $4(1-x) = 1$, which simplifies to $-4x = -3$ or $x = 3/4$.

There is a second fixed point at $x = 3/4$. So far so good: We have found two very important orbits.

Now consider the orbit of $x_0 = 0.5$. We have

$$0.5 \to 1 \to 0 \to 0 \to \cdots$$

This is an example of an **eventually fixed** orbit: The point $x_0 = 0.5$ is not itself fixed, but its second iterate is fixed. Again we have found a very special orbit.

Now let's try a few more. The chart on the next page shows the first 25 points (listed to 5 decimal places) on the orbits of the seeds $x_0 = 0.1$, $x_0 = 0.123$, and $x_0 = 0.6$.

Seeds:	0.1	0.123	0.6
0	0.1	0.123	0.6
1	0.36	0.43148	0.96
2	0.9216	0.98122	0.1536
3	0.28901	0.0737	0.52003
4	0.82194	0.27308	0.9984
5	0.58542	0.79402	0.00641
6	0.97081	0.65421	0.02547
7	0.11334	0.90488	0.09927
8	0.40197	0.34428	0.35767
9	0.96156	0.90301	0.91897
10	0.14784	0.35033	0.29786
11	0.50392	0.9104	0.83656
12	0.99994	0.3263	0.54692

Seeds:	0.1	0.123	0.6
13	0.00025	0.87931	0.9912
14	0.00098	0.4245	0.03491
15	0.00394	0.9772	0.13476
16	0.01568	0.08913	0.4664
17	0.06174	0.32474	0.99548
18	0.23173	0.87714	0.01798
19	0.71212	0.43106	0.07062
20	0.82001	0.98099	0.26253
21	0.59036	0.07461	0.77444
22	0.96734	0.27616	0.69873
23	0.12638	0.79958	0.84203
24	0.44165	0.641	0.53206
25	0.98638	0.92048	0.99589

Do you see the pattern here? Definitely not! These orbits do not appear to be tending to a fixed point or cycling. Instead, they seem to jump all over the place and to behave very differently from orbits of linear iterations. They seem chaotic. Perhaps we will see a pattern if we view these orbits as time-series graphs or use graphical iteration.

TIME-SERIES GRAPHS

The orbits listed above seem to be wandering around the interval $0 \leq x \leq 1$ rather aimlessly. Let's see if we can detect a pattern from the time series for one of these orbits. Here is the time-series graph for the seed $x_0 = 0.123$, consisting of the first 100 points on the orbit:

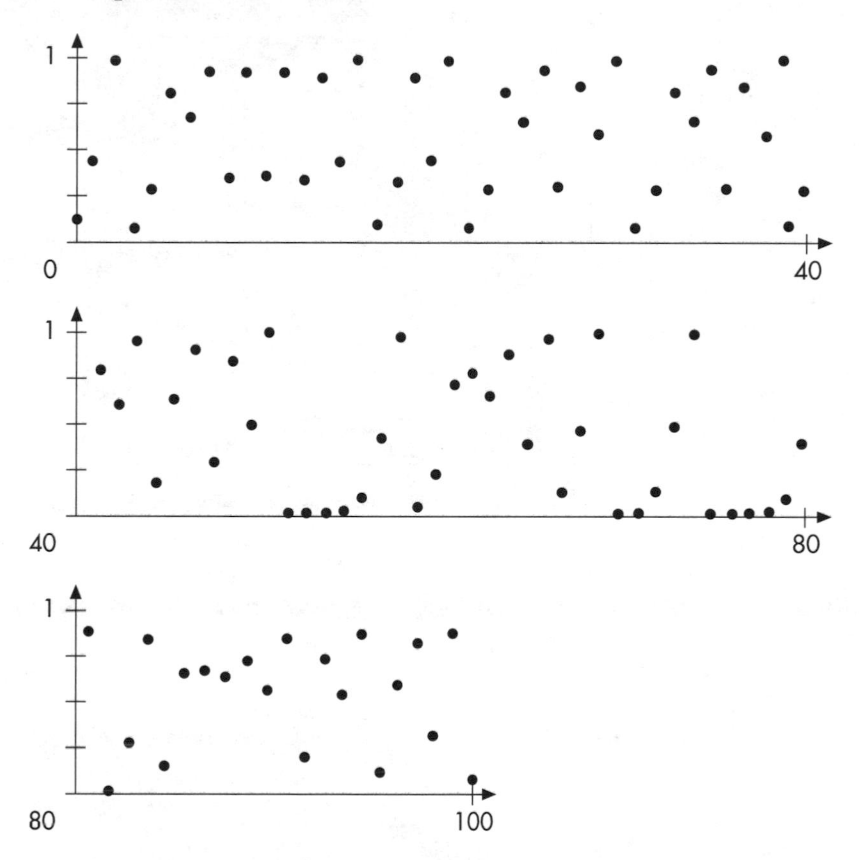

Just when you think you are beginning to see a pattern in this picture, the time-series graph begins to do something else and a new pattern emerges. A few iterations later, however, this pattern also begins to disintegrate and something else appears. There does not seem to be any overall pattern in this picture. This is our first glimpse of what mathematicians call **chaos**.

GRAPHICAL ITERATION

Let's try that again, this time using graphical iteration. Here is a picture of the first 100 points on the orbit of 0.123 under the rule $x \to 4x(1 - x)$:

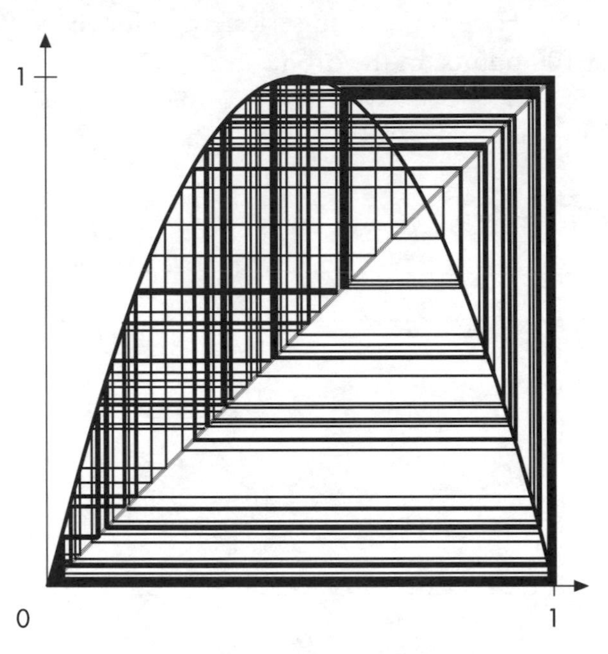

Again, there does not seem to be any pattern here. Here is a similar graphical iteration, this time for the seed 0.2:

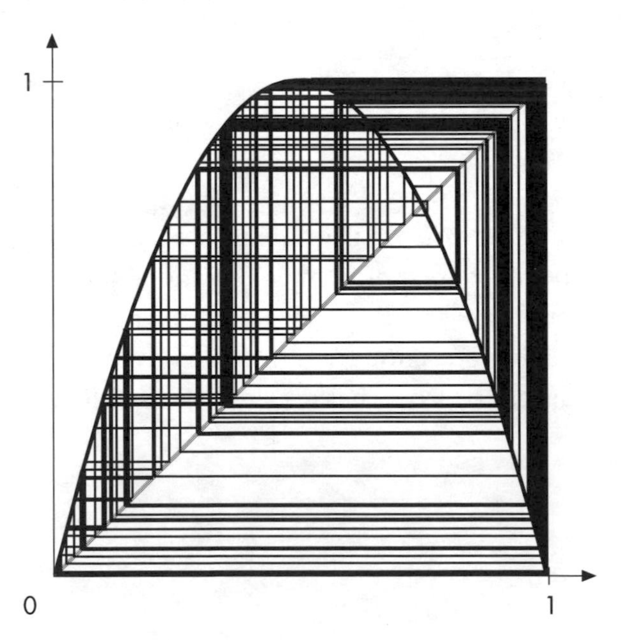

Again, no help. It appears that orbits wander around without any discernible pattern.

Watch out! Sometimes graphical iteration can be deceiving. For example, here is the graphical iteration showing the first 125 points on the orbit of 0.1167 under the iteration rule $x \to 3.84x(1 - x)$:

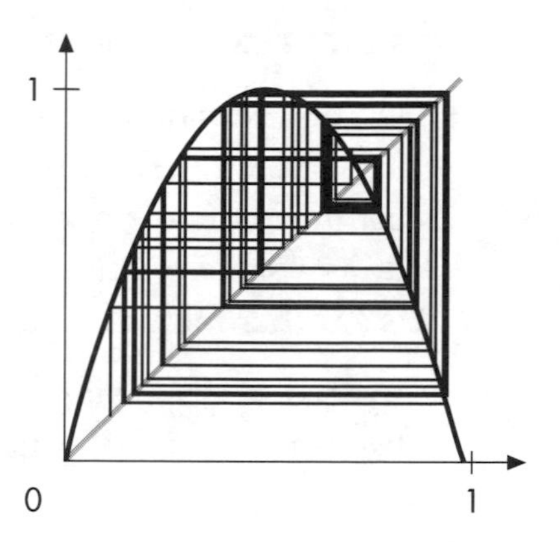

At first glance, this orbit looks somewhat chaotic. However, if we plot only iterations 100, 101, . . . , 125, we find something quite different:

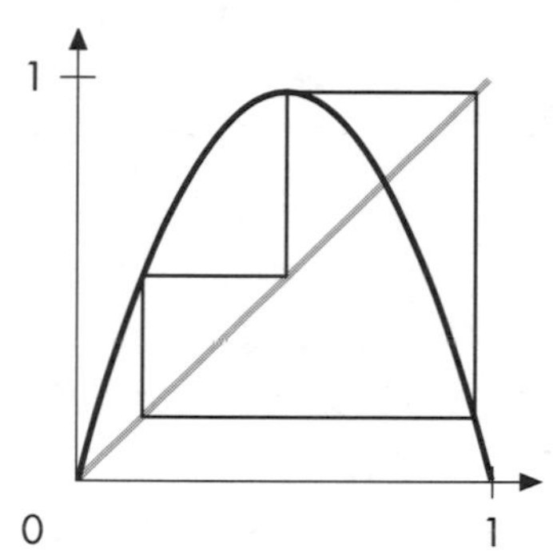

The orbit has in fact tended to a 3-cycle. However, it has taken a while before this orbit settles down on this 3-cycle. What we saw in the first graphical iteration was transient behavior. The orbit hops around for a while before being attracted to the 3-cycle.

The time-series graph shows this more clearly. After approximately 40 iterations, the orbit settles down on a 3-cycle:

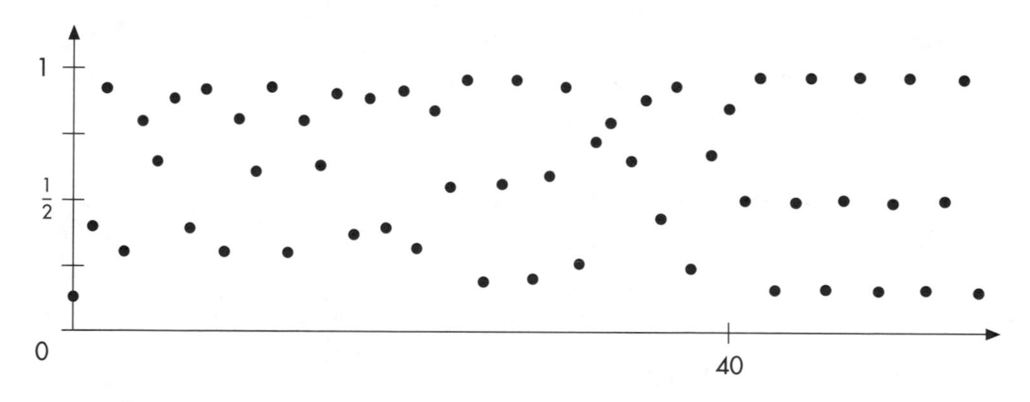

HISTOGRAMS

Do not despair. At this point, you might begin to wonder if we can ever understand iterations that do not eventually settle down. Curiously, while the list of numbers on such an orbit, the time-series graph, and graphical iteration do not seem to give us any useful information about the fate of the orbit, there is another picture that sometimes exhibits a pattern.

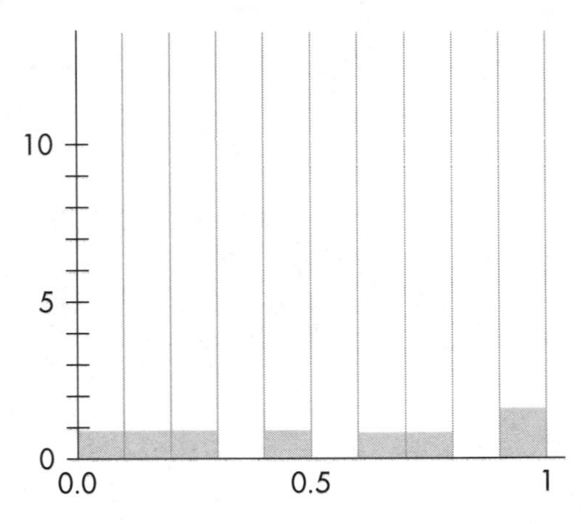

Instead of displaying the orbit as a simple list of numbers, let's make a **histogram** of an orbit of $x \to 4x(1 - x)$. To do this, first break up the interval $0 \leq x_0 \leq 1$ into a collection of equal-sized intervals. For example, at the outset, suppose we break up the interval $0 \leq x_0 \leq 1$ into smaller intervals of length 0.1. That is, think of this interval as being subdivided into the tiny intervals $0.0 \leq x < 0.1$, $0.1 \leq x < 0.2$, $0.2 \leq x < 0.3$, and so forth. Now every time the orbit of the seed enters one of these intervals, place a unit rectangular block in a stack over that interval. The first eight points on the orbit of the seed 0.123 are 0.123, 0.431, 0.981, 0.073, 0.273, 0.794, 0.654, and 0.905.

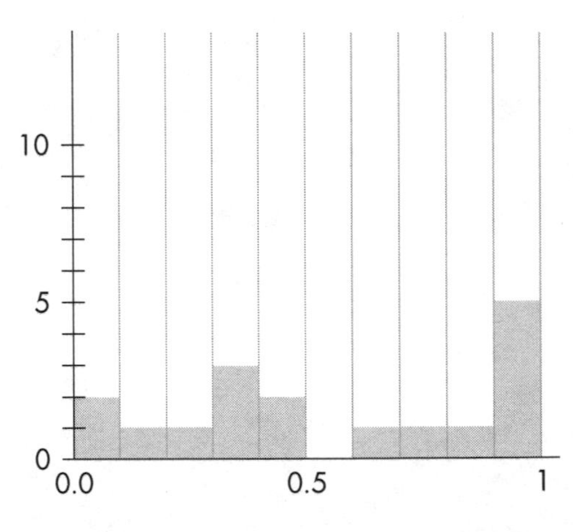

So we place two blocks over the interval $0.9 \leq x < 1.0$ (corresponding to the two points on the orbit 0.905 and 0.981 that lie in this interval). We put one block in $0.0 \leq x < 0.1$ corresponding to 0.073, one block in $0.1 \leq x < 0.2$ corresponding to 0.123, and so forth. Adding the next nine points to the orbit yields 0.123, 0.431, 0.981, 0.073, 0.273, 0.794, 0.654, 0.905, 0.344, 0.903, 0.350, 0.910, 0.326, 0.879, 0.425, 0.977, and 0.089.

Usually, when we compute a histogram of an orbit, we choose a much finer division of the axis as well as many more points on the orbit. For example, in the histogram of the orbit of 0.123 under $x \to 4x(1 - x)$, shown at right, we have broken up the interval into 400 equal-sized pieces (each of length 0.0025) and computed 10,000 iterations.

In this histogram, note that there are many more points congregated near 0 and 1, and fewer in the middle. Now here comes the surprise. Let's compute the histogram for the same rule, but now using the seed 0.1. Recall that this orbit differed completely from that of 0.123 after just a few iterations.

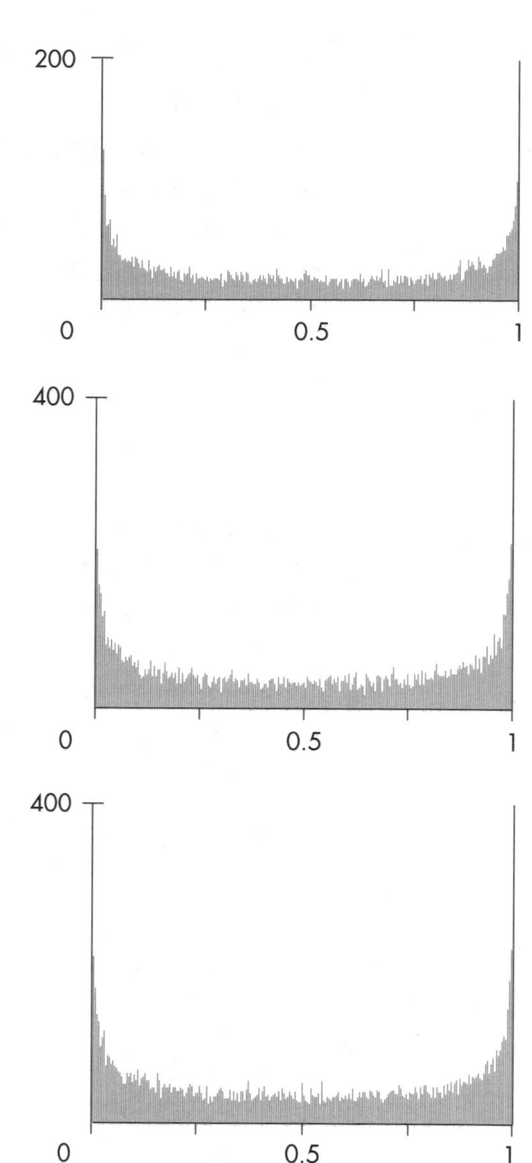

But the histograms are essentially the same—peaks at 0 and 1 and a big valley in between. This is what we meant when we said that the orbits may be very different, but the *pictures* of the orbits are almost identical. The third histogram here is for the seed 0.6 under the same iteration rule.

Again we see a valley between two peaks. There are small differences between any two of the histograms, but their essential structure remains the same.

Random seeds

To get these particular histograms, we chose a seed "at random." That is, we carefully avoided choosing special seeds such as 0 or $3/4$. (Recall that these were the fixed points for this iteration rule.) Had we chosen these points to display the histogram, we would have found a much different picture, one with just a single spike over the chosen fixed point. In general, there is no way to determine ahead of time which seeds will produce distinctive histograms. However, it is a fact that almost all seeds yield histograms like those shown above.

THE CRITICAL ORBIT

The point on the horizontal axis where the graph of the logistic iteration rule reaches a maximum or minimum value is called the **critical point**. For example, the graph of $y = kx(1 - x)$ reaches a maximum value at $x = 0.5$. That maximum value is

$$k(0.5)(1 - 0.5) = \frac{k}{4}$$

The maximum value when $k = 4$ is 1, and that maximum is taken when $x = 0.5$. So we say 0.5 is the critical point for the rule $x \to kx(1 - x)$. Another way to say this is that the graph of the rule has slope 0 at the point on the graph directly over the critical point.

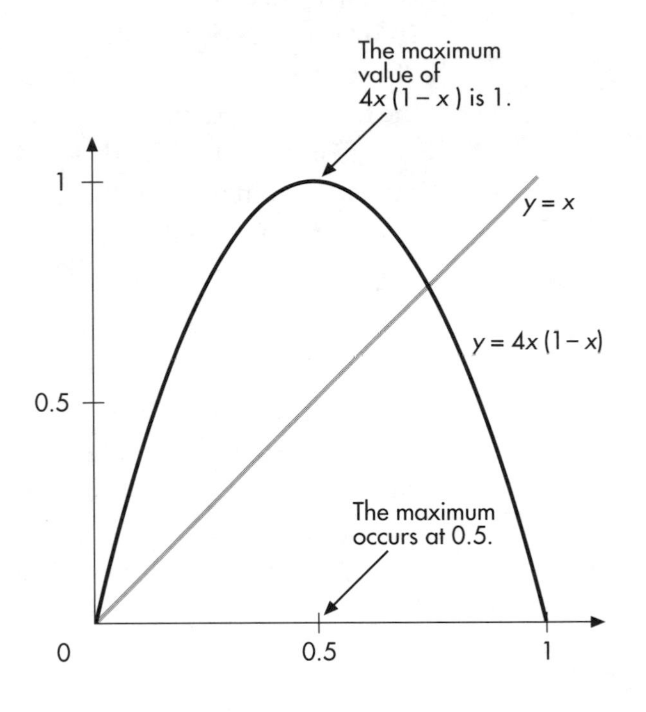

The maximum value occurs at 0.5 since we can rewrite $kx(1 - x)$ as

$$kx(1 - x) = -k\left(x^2 - x\right)$$

Completing the square on the right-hand side yields

$$\begin{aligned}-k\left(x^2 - x\right) &= -k\left(x^2 - x + \frac{1}{4}\right) + \frac{k}{4} \\ &= -k\left(x - \frac{1}{2}\right)^2 + \frac{k}{4}\end{aligned}$$

and the largest this sum can be is $k/4$, and that happens when $x = 1/2$, or 0.5.

Remember that in the special case $x \to 4x(1 - x)$, the orbit of 0.5 (the **critical orbit**) is eventually fixed. This orbit is

$$0.5 \to 1 \to 0 \to 0 \to \cdots$$

Notice what has happened. The forward part of this orbit (everything after 0.5) consists of just two numbers, 0 and 1. Interestingly, these are exactly the places where the histogram for $4x(1 - x)$ has peaks.

A DIFFERENT ITERATION RULE

Is this a coincidence? Suppose we change the iteration rule to $x \to 3.9x(1 - x)$. Graphical iteration shows that the orbits appear chaotic. (We will describe what "chaos" means more precisely in the next lesson. For now let's just observe what happens to orbits.) Here are the first 125 points on the orbit of 0.1 and 0.2 under this rule:

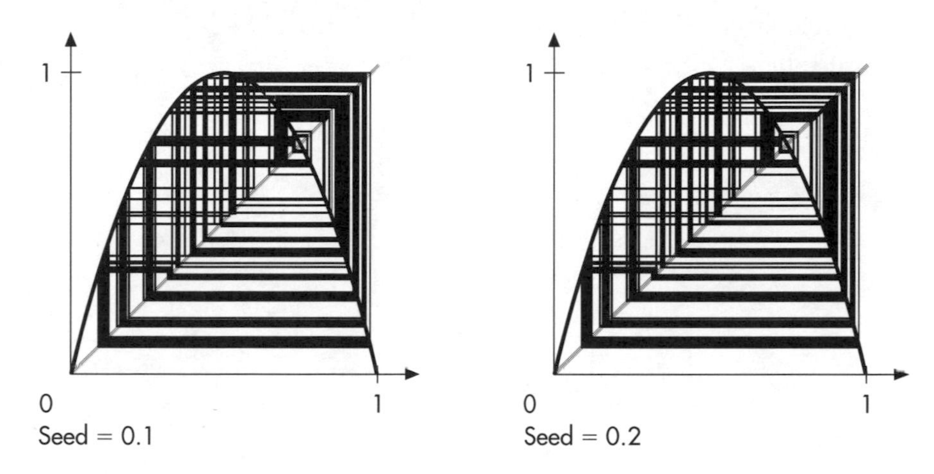

But the histograms of orbits are essentially the same. Here are the histograms for 0.1 and 0.2:

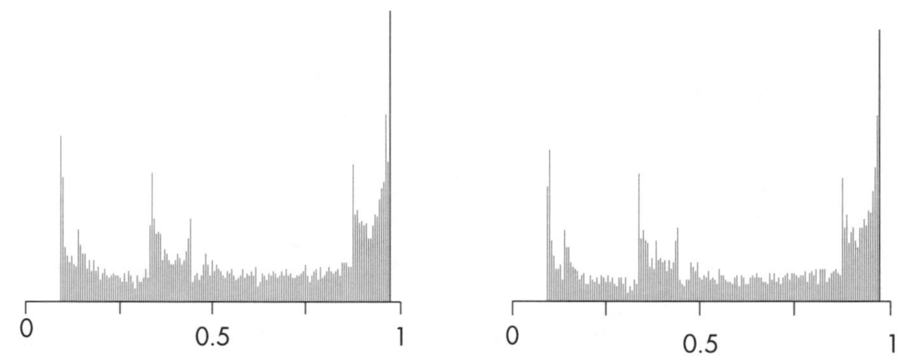

This time the histograms differ from the previous rule, but both seeds generate essentially the same picture. Note how the various peaks and valleys resemble one another. So, while the details of a chaotic system may be completely unpredictable, when we put everything together we see a pattern amid the chaos.

Now consider the orbit of 0.5. This orbit is

$$0.5 \to 0.975 \to 0.095 \to \cdots \to 0.336 \to \cdots \to 0.869 \to \cdots \to 0.443 \to \cdots$$

Now look at the locations of the peaks of the histogram at right. The peaks occur exactly over the forward part of the critical orbit (everything beyond 0.5)!

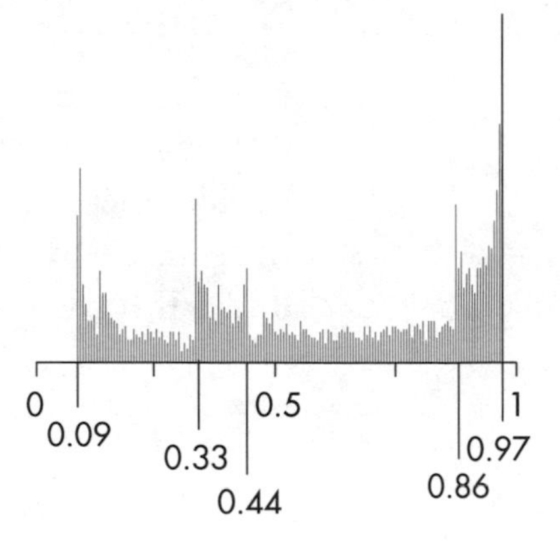

In later lessons we will see that there is even more fascination associated with the critical orbit. In a very real sense, the fate of the orbit of the critical point tells us a lot about the fate of all other orbits.

NAME(S):

1 ▹ CONSTRUCTING HISTOGRAMS

a. Construct a histogram for the iteration rule $x \to 4x(1 - x)$ using the first 50 points on the orbit of 0.3 and 0.9.

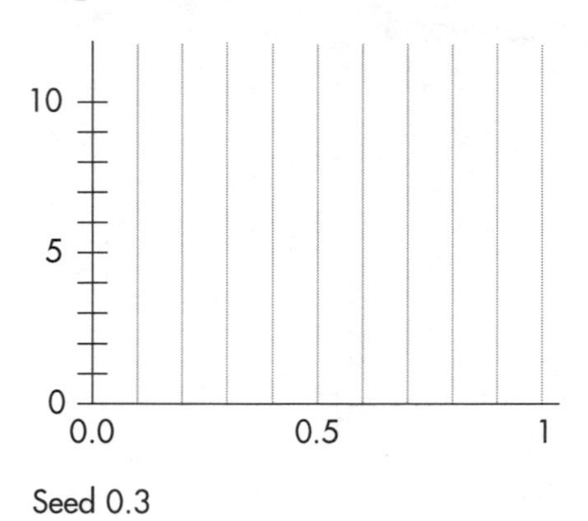

Seed 0.3

10
5
0
0.0
0.5
1

Seed 0.9

b. Compare your results. Do you begin to see the same peaks and valleys as before?

____

____

____

____

COMPUTER INVESTIGATION

c. Use a computer or spreadsheet to compute the next several hundred points on the orbit. Do the peaks and valleys emerge now?

____

____

____

____

2 ▹ SIMPLER HISTOGRAMS

a. Not all iteration rules yield chaotic results, so not all histograms look the same. For each of the following iteration rules, compute the first 30 points on the orbit of the given seed and plot the corresponding histogram.

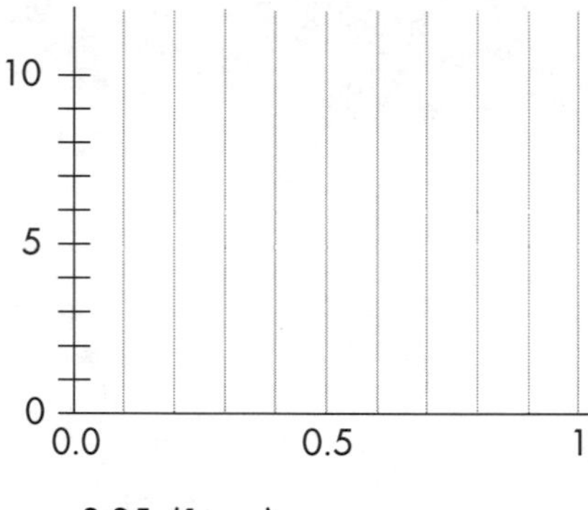

$x \rightarrow 3.25x(1 - x)$
Seed 0.25

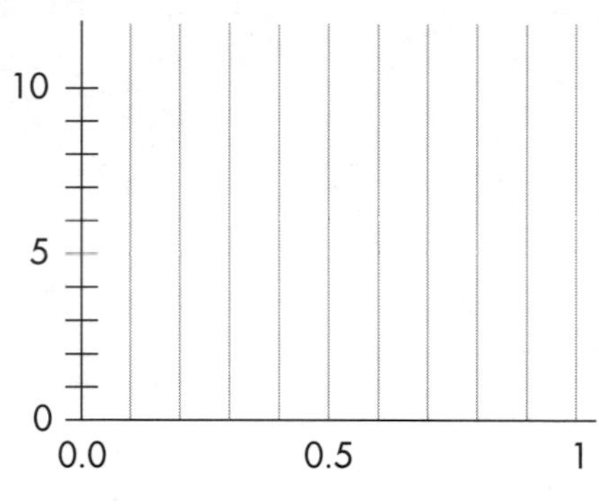

$x \rightarrow 3.84x(1 - x)$
Seed 0.4

b. Discuss the notion of transient behavior for the two rules in part a. What happens if we consider more iterations? How does the histogram change?

___

___

___

3 ▹ Time series and histograms

Here is a time series for an iteration rule. Plot the corresponding histogram (as accurately as possible) using intervals of length 0.1. Here we have "connected the dots" in the time-series graphs to help you follow the orbit:

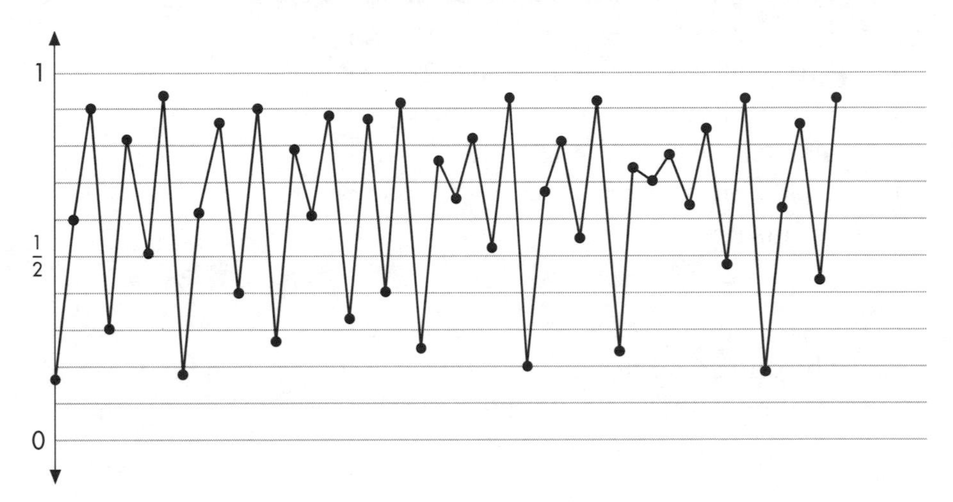

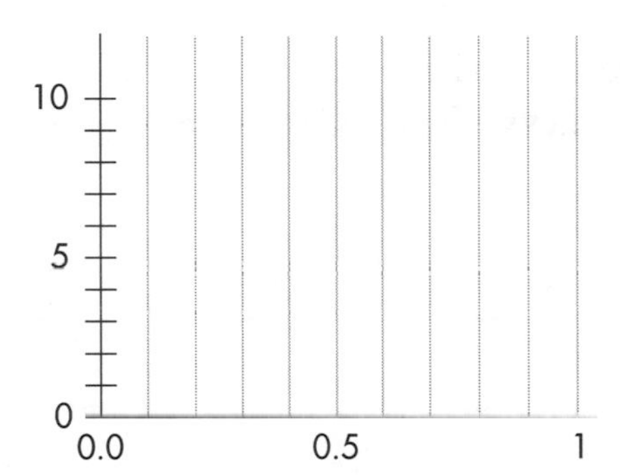

4 ▹ A QUADRATIC HISTOGRAM

COMPUTER INVESTIGATION

a. Use a computer or spreadsheet to compute a histogram for the orbit of 0.1 under the iteration rule $x \to x^2 - 2$. In this case, your orbit will live in the interval $-2 < x < 2$, so you should break up this interval into 40 subintervals of length 0.1. Use at least 2000 points on the orbit. Sketch your results here.

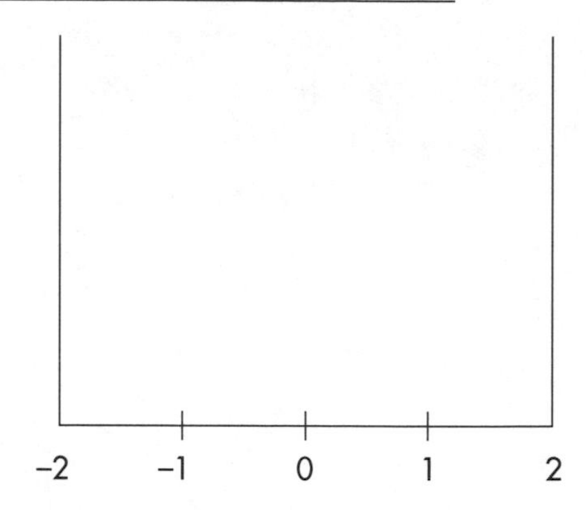

b. The critical point for $y = x^2 - 2$ is 0, since the graph of $y = x^2 - 2$ has a minimum value there (and so the slope is 0). Compute the forward orbit of 0 under this iteration rule.

____

c. What can you say about the relationship between this forward orbit and the peaks and valleys of your histogram?

____

____

____

____

5 ▹ MATCHING HISTOGRAMS AND GRAPHICAL ITERATIONS

Here are four histograms and on the following page four graphs of logistic iteration rules $x \rightarrow kx(1 - x)$ for different k-values. Each graph displays the first several points on the orbit of the critical point, 0.5. Match the histograms and graphs corresponding to the same k-value.

A. 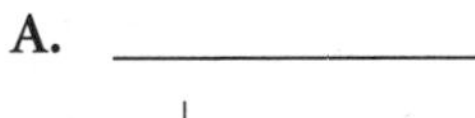

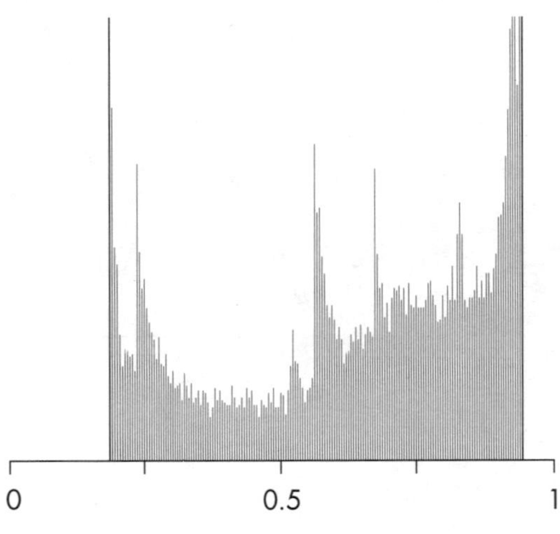

B. 

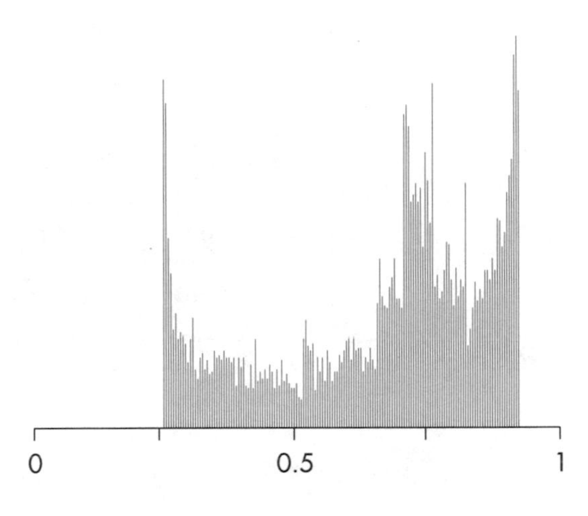

C. ______________

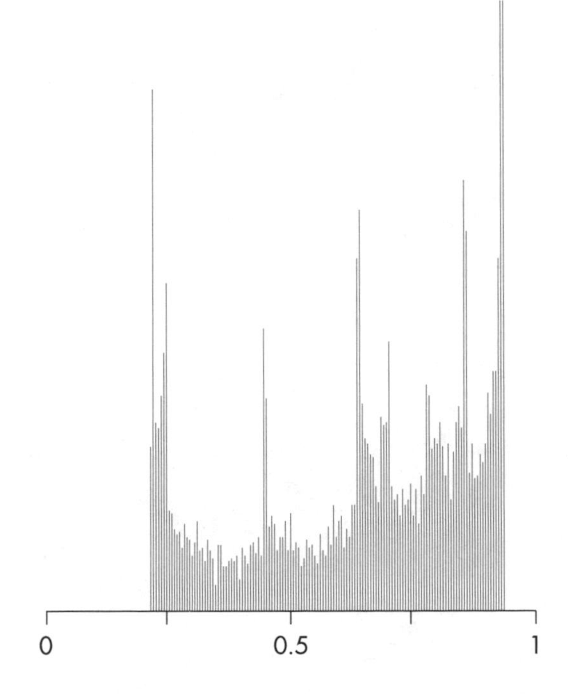

D. ______________

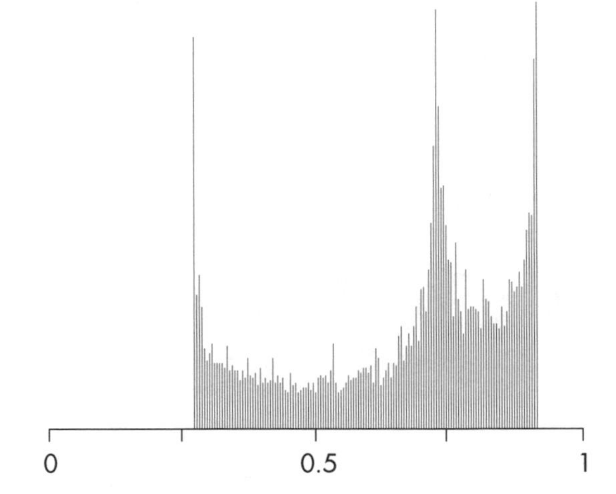

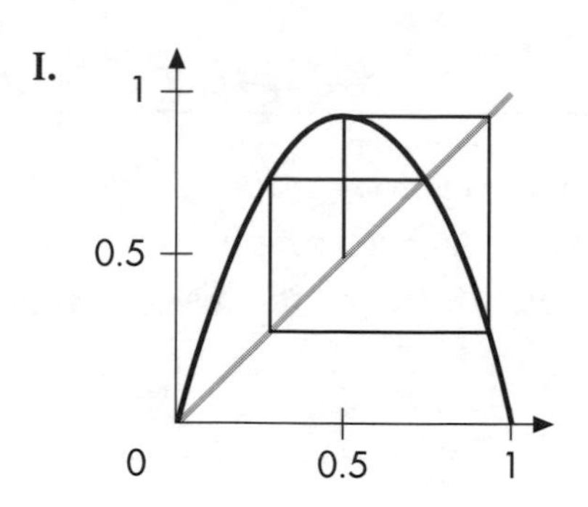
I.
1
0.5
0
0.5
1

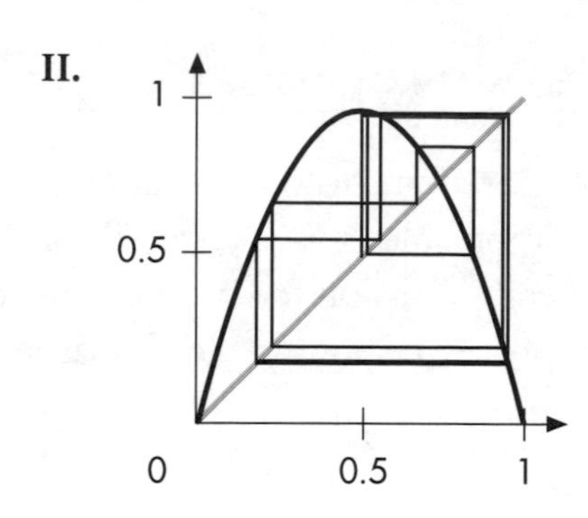
II.
1
0.5
0
0.5
1

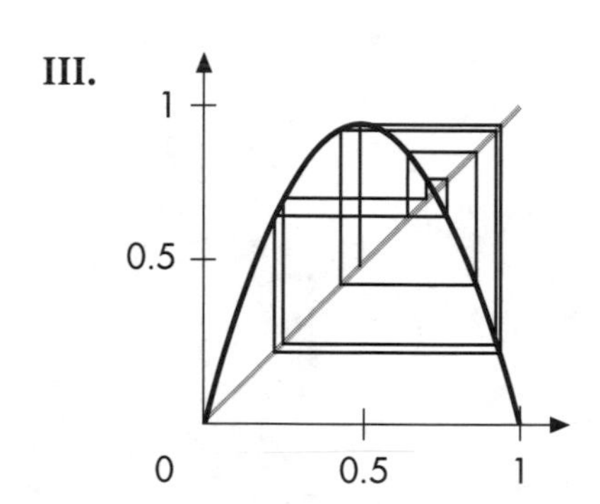
III.
1
0.5
0
0.5
1

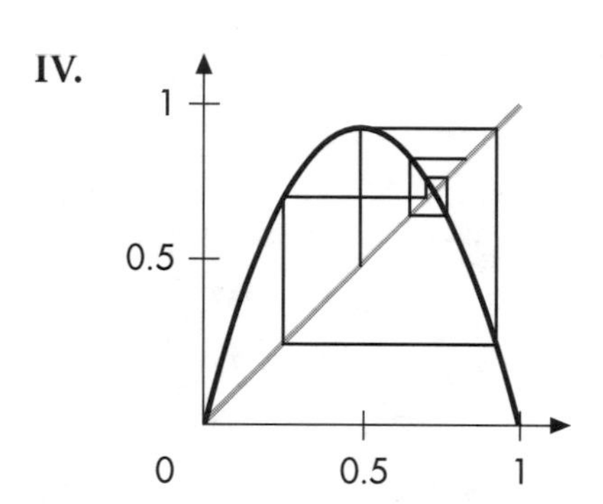
IV.
1
0.5
0
0.5
1

NAME(S):

1. Compute the histogram for the logistic iteration rule $x \to 3.83x(1 - x)$ first using the seed $x_0 = 0.5$, then using the seed $x_0 = 0.1$. Do these two orbits share the same fate? Which orbit yields this fate faster?

2. Compute the histogram for the logistic iteration rule $x \to 3.739x(1 - x)$ first using the seed $x_0 = 0.5$, then using the seed $x_0 = 0.1$. Do these two orbits share the same fate? Which orbit yields this fate faster?

Teacher Notes

The Butterfly Effect

Overview

This lesson takes a deeper look at chaotic behavior and its unpredictability by investigating a phenomenon known as "sensitivity to initial conditions." This is the main ingredient of chaos, and it has important ramifications in science as well as mathematics.

Mathematical Prerequisites

Only a basic knowledge of iteration is necessary to understand this lesson. Students should know that two numbers are close if their decimal expansions agree in the first several places.

Mathematical Connections

One of the major impacts of this material in the mathematics classroom is creating an awareness of the **limitations of the accuracy of calculators and computers** due to the number of decimal places that are carried internally. Viewing sensitivity to initial conditions improves students' skills at **estimation** and helps them acquire a stronger sense of "how close is close" since they are forced to look at columns of numbers and determine when these numbers become "far apart." The Investigations in this lesson carry an important message for students whose favorite expression is "my answer is close enough."

Technology

Calculators or spreadsheets are more than sufficient to view sensitive dependence. It might prove interesting to have students performing the same investigation on calculators that carry different numbers of decimal places, or to have them do identical investigations on different brands of calculators to determine whether the two brands have the same decimal accuracy.

SUGGESTED LESSON PLAN

CLASS TIME

One 50-minute period is enough if students are conducting investigations in class to observe sensitivity to initial conditions. Another option is to assign this material as outside reading, since much of the Explanation deals with history, applications, and even a little philosophy.

PREPARATION

Plan to demonstrate sensitivity with several nearby seeds for the iteration rule $4x(1 - x)$. Check beforehand that you actually do find sensitivity. If your seeds agree to too many decimal places, the calculator will cut off the final digits and you will effectively compute the same orbit in each case.

The historical material in this section may be supplemented by material from James Gleick's *Chaos: Making a New Science* (New York: Viking Penguin, 1987).

LESSON DEVELOPMENT

There are several different ways to approach this lesson. One method is to have students read the Explanation ahead of time and complete Investigations 1 and 2 prior to class. Students could share their results in class and discuss the ideas from the lesson, then complete Investigations 4 and 5 in class in cooperative groups, perhaps using spreadsheets to speed up the computations. An alternative approach is to begin the lesson by having students perform some of the Investigations in class, summarize the results, then read the Explanation for homework.

The material in this lesson is an ideal topic for students to report on chaos. Ask them to investigate several different iteration rules and determine which yield sensitivity and which do not. They should prepare an essay complete with graphs and tables explaining their conclusion. Alternatively, ask students to report on real-world examples of sensitivity to initial conditions as in Investigation 6.

Investigation 2 is a pencil-and-paper method of illustrating sensitivity and also of reinforcing the mechanics of graphical iteration. You can use the transparency master that follows to accomplish this in class and have students compare their results; students could record their results directly on transparencies and then overlay several of them to observe any differences in the orbits. Make sure they understand why the fact that different students produce different results demonstrates sensitivity to initial conditions if they supposedly all started with the same seed.

LESSON NOTES

Be careful with Further Exploration problem 2! Depending on which machine you use, you may get radically different answers. For example, most computers will say that the orbit eventually becomes fixed at 0. Some machines, however, actually do integer arithmetic and give correct answers. The fact is that the doubling function is quite sensitive to initial conditions.

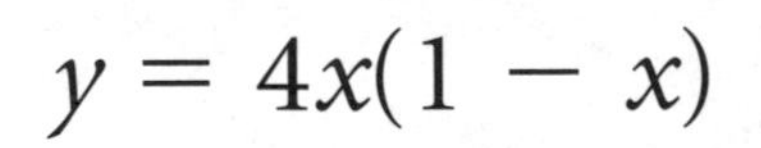

$y = 4x(1 - x)$

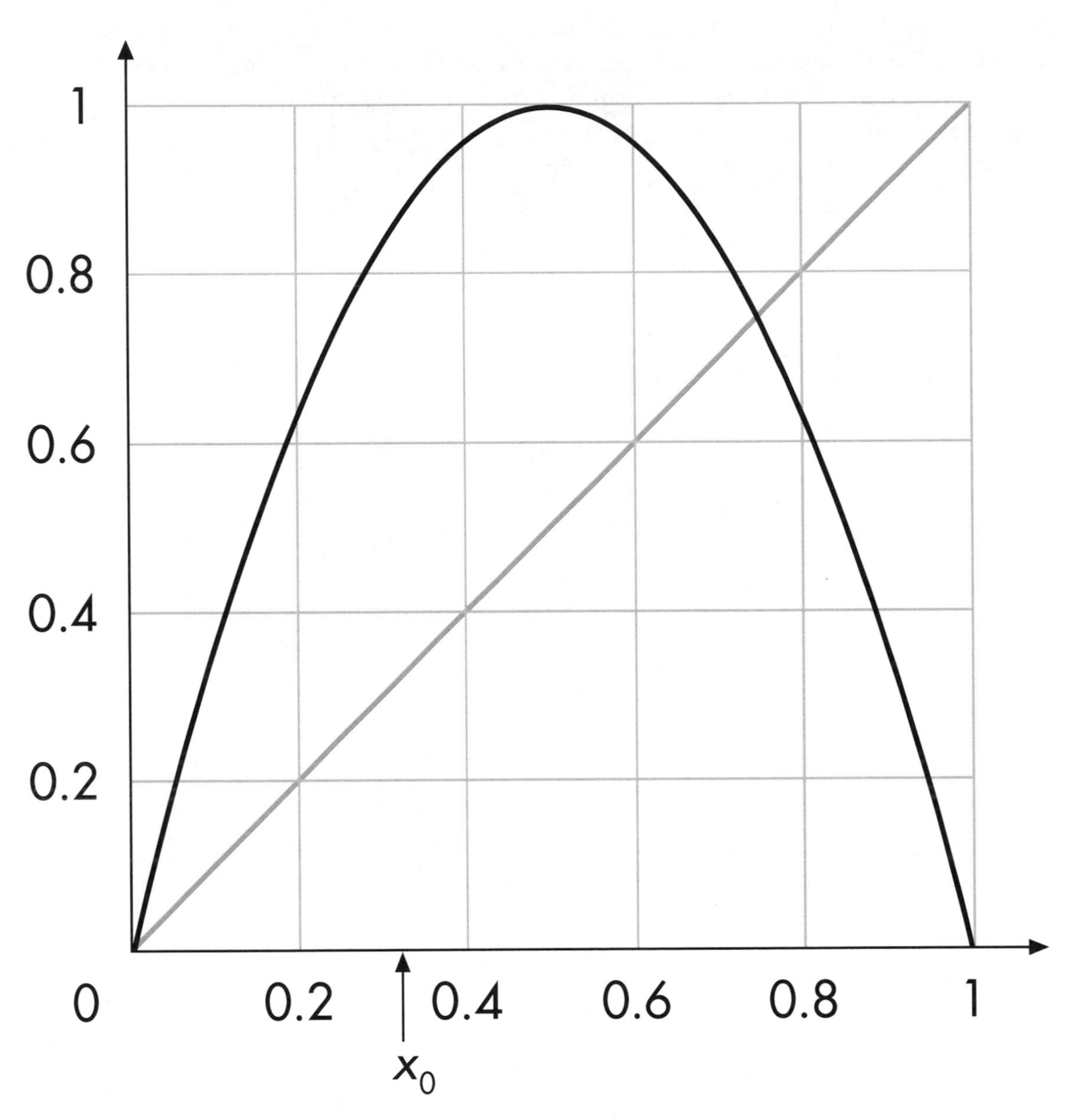

In the previous lesson, we saw that the logistic iteration rule sometimes yields orbits that behave quite unpredictably. In this lesson, we will examine this unpredictability even further and see that the possibility of a system behaving chaotically has ramifications in all areas of science.

SENSITIVITY TO INITIAL CONDITIONS

If you look at the logistic iteration rule $x \to 4x(1 - x)$ beginning with the seed 0.5, you find a very simple orbit, namely, the eventually fixed orbit

$$0.5 \to 1 \to 0 \to 0 \to \cdots$$

Now let's change the seed slightly, say, to $x_0 = 0.50001$. So the new seed differs from the old seed only in the fifth decimal place. But look what a difference this makes in the fate of the orbit. Here is the time-series graph for the first 35 points on the orbit of both seeds (0.5 unshaded). For the first 13 or so iterations, the orbit of 0.50001 is very close to the orbit of 0.5 (the shaded dots are covering the unshaded dots). But then the orbit of 0.50001 begins to move away from the eventually fixed orbit. Then it seems to assume a life of its own and bears no resemblance to the orbit of its eventually fixed neighbor.

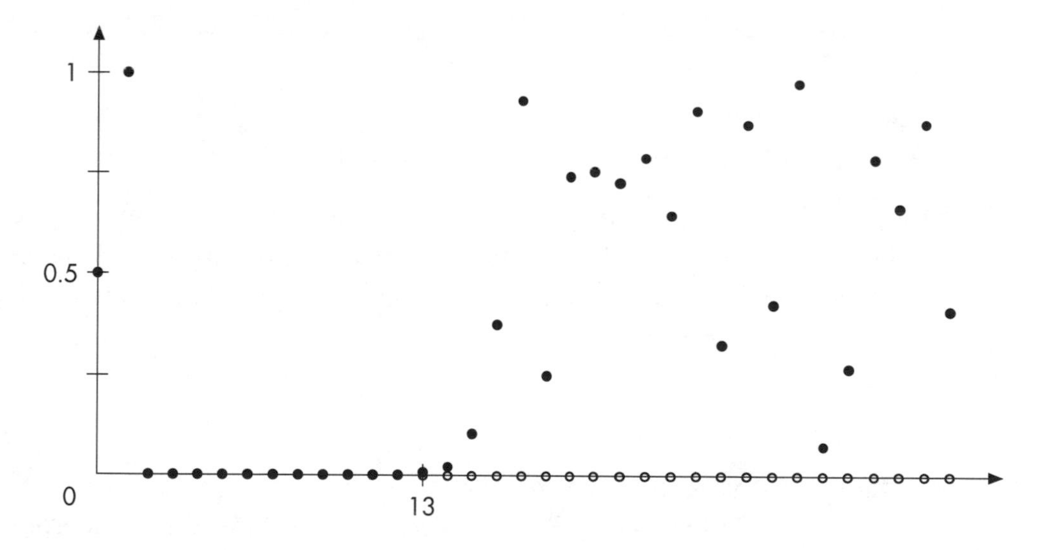

Two nearby seeds have vastly different fates. After a very few iterations, the orbits move apart and their remaining orbits behave very differently. This phenomenon is called **sensitivity to initial conditions.**

Here is another example. Consider the seeds $x_0 = 0.123$ (shaded) and $x_0^* = 0.12301$ (unshaded). Again these seeds are very close together. Their orbits initially are very close too. For the first few iterations, the shaded dots hide the unshaded dots because they are so close together. But look what happens after just 12 iterations: Their time-series graphs begin to separate from each other. By the sixteenth iteration, the orbits are very far apart and eventually behave very differently.

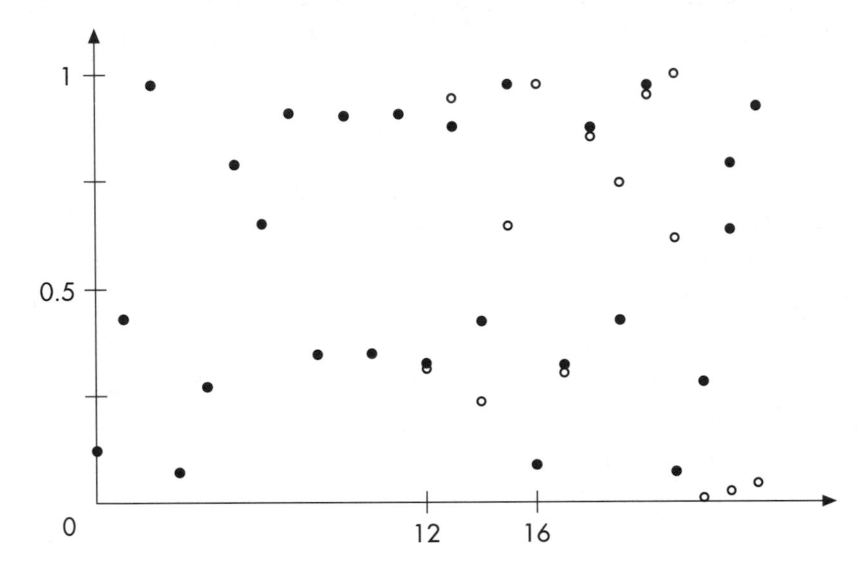

Again we see that nearby orbits separate from one another, and so we have sensitivity to initial conditions.

This sensitivity is typical in a system that is chaotic. An iteration rule is called **chaotic** on an interval if *all* seeds in the interval exhibit this kind of sensitivity. As we have begun to see above, the iteration rule $x \rightarrow 4x(1 - x)$ is sensitive to initial conditions on the interval $0 \leq x_0 \leq 1$.

THE BUTTERFLY EFFECT

The fact that simple iteration rules can be so sensitive to initial conditions was discovered only recently. Back in the 1960s, the American meteorologist Edward N. Lorenz was using the kind of mathematics that you are now studying to try to understand a highly simplified model of the weather. This was the period of time when very crude computers were just beginning to be used in science and mathematics. Lorenz was using his computer to plot a time-series graph of particular orbits of his weather model and noting, just as you did above, that many of the time-series graphs had no pattern. One day he received a phone call while the computer was producing a particular time-series graph, so he had to leave the computation running while he answered the phone. When he returned to the computer, he found only the tail end of the time series, so he decided to repeat the calculation to see the entire plot. However, he forgot that he had entered a six-digit decimal and entered only the first four digits. So his new seed was off in the fifth decimal place. When he computed the new time-series graph, he found that the tail of the plot bore no resemblance to the original plot. Lorenz had discovered sensitivity to initial conditions.

Lorenz's discovery has had a profound effect on all sciences, not just meteorology. Lorenz reasoned that if a small change in a far-off decimal place could completely change the output of his model, then there was little chance that he could ever use his model to predict the weather. Indeed, the weather is a much more complicated system than our simple iterative systems. If these simple models are so sensitive to initial conditions, perhaps the weather is as well. The idea is that if a butterfly flaps its wings in a distant place such as Brazil, then that will not have a tremendous effect on the weather in Brazil right away. But as the ripples in the air caused by the slight flapping of the wings change the various wind speeds and directions (ever so slightly at first), this could set in motion a chain of events that will radically alter the world's weather sometime much later. This is real sensitivity to initial conditions!

Nobody has ever verified the fact that a single butterfly can change the world's weather, but the fact remains that if our simple mathematical models can have such tremendous sensitivity, why should the intricate models that arise in nature not be even more complicated?

Other applications

Our simple model for population growth exhibits, for certain parameter values, sensitivity to initial conditions. It is reasonable to expect that more complicated population models that build on the hypotheses we adopted in the logistic model should also exhibit the same type of behavior.

In the years since Lorenz's discovery, scientists in all disciplines from astronomy to zoology have reexamined their mathematical models for instances of chaotic behavior. Very often they have found that in fact this chaos was present all along. In the past, for example, cardiologists often thought that the healthy human heart is one that beats rhythmically. This turns out to be false: EKG data from healthy hearts often show small fluctuations from "regular," whereas certain individuals with cardiac arrhythmia can have EKG data that are completely regular, like a time-series graph exhibiting cyclic behavior.

In another example, astronomers always believed that the motion of the planet Pluto is quite regular, but recent studies have shown that this is not the case. The orbit does not trace an ellipse but rather, wanders around in a more complicated fashion in its area of the solar system, another indication of chaos.

Computing chaotic systems

The fact that simple iteration rules can exhibit sensitivity to initial conditions throws into doubt any numerical computations done on such systems. Here is the reason. Whenever a computer makes computations, it inevitably makes small errors. This can be caused by roundoff or other forms of "truncation." Obviously, the computer cannot store numbers with infinite accuracy. In fact, the computer stores numbers with only a relatively small degree of accuracy. Depending on which machine you are using, you may be able to store numbers with only 8 or 16 decimal places of accuracy. Now, if you multiply two such numbers, generally you will double the number of decimal places. For example, if you multiply 0.11111111 by 0.22222222, both with 8 decimal places, you end up with 0.246913575308642, which has 15 decimal places. (You probably should check that this answer is correct, since the limitations of accuracy of our favorite calculator forced us to do this computation by hand.) Our calculator did not store enough digits to make the computation with complete accuracy.

When confronted by a number that is "too long," the computer cuts off, or truncates, the tail of the number and makes an appropriate roundoff. How and when the computer does this varies from machine to machine. So, for example, our computer tells us that the product of 0.11111111 and 0.22222222 is 0.24691358 instead of the 15-digit number above. That's a pretty accurate answer for most purposes, but think of what would happen if this computation arose while we were computing an orbit of a chaotic iteration rule. We would have made a tiny error in the eighth decimal place. No big deal at this iteration. But in subsequent iterations, this tiny error becomes magnified. Each time we iterate, a new and perhaps larger error is introduced into the computation. These errors are compounded as we continue to iterate, and eventually there is no similarity between the real orbit we were computing and the numbers the computer is displaying. That is not a happy situation.

EFFECTS OF ROUNDOFF

Let's investigate the butterfly effect by iterating the rule $x \rightarrow 4x(1 - x)$ using some crude and useless computers. The first computer is so bad that it retains only the first decimal place. The second is a little better, but still pretty bad, as it retains two decimal places of accuracy. In each case, let's compute the orbit of 0.1 using these computers. In the first case, we find

$$x_0 = 0.1$$

$$x_1 = 4 \cdot 0.1(1 - 0.1) = 0.36 \approx 0.4$$

(remember that we keep only one decimal place of accuracy with this rule)

$$x_2 = 4 \cdot 0.4(1 - 0.4) = 0.96 \approx 1.0$$

$$x_3 = 4 \cdot 1.0(1 - 1.0) = 0$$

and 0 is a fixed point. So after three iterations using this inept computer, this orbit is eventually fixed.

If we keep two decimal places of accuracy, we find

$$x_0 = 0.1$$

$$x_1 = 4 \cdot 0.1(1 - 0.1) = 0.36 \text{ (now we keep two decimal places)}$$

$$x_2 = 4 \cdot 0.36(1 - 0.36) = 0.9216 \approx 0.92$$

$$x_3 = 4 \cdot 0.92(1 - 0.92) = 0.2944 \approx 0.29$$

$$x_4 = 4 \cdot 0.29(1 - 0.29) = 0.8236 \approx 0.82$$

$$x_5 = 4 \cdot 0.82(1 - 0.82) = 0.5904 \approx 0.59$$

Obviously, we do not have an eventually fixed orbit at this stage, so already these two computers have produced differing results.

The computer you use is better, but not that much better. It may retain 8 decimal places, or maybe 16 decimal places. But still the computer is making roundoff errors, so the phenomenon we see with the crude computers above still persists.

Unpredictability

An iteration rule that is sensitive to initial conditions is essentially unpredictable. In order to make predictions, we must know our seed with infinite precision, which is impossible. Even if we know that we have only finitely many decimal places, the errors introduced by truncation affect the result as well. So we either have to carry more and more decimal places as we iterate, or we have to give up any semblance of accuracy. As we saw earlier, the number of decimal places grows very quickly, so carrying out completely accurate computations is again impossible in a practical sense.

This is one of the big discoveries in modern mathematics, that very simple iteration rules can be completely unpredictable. This result applies to much more than just iteration rules. Other kinds of mathematical models that you will encounter in calculus (such as differential equations) are now known to exhibit sensitivity as well.

The patterns in chaos

Now let's return to some of the ideas of the previous lesson. As we have seen, when a system is sensitive to initial conditions, the list of numbers produced by the computer can be completely inaccurate. On the other hand, when we plot these numbers graphically in a histogram, we almost always find the same picture. Can we trust this image? Mathematicians have proved that most orbits in the kinds of chaotic systems we have encountered actually do yield the "right" pictures. What the word *right* means is difficult to explain without calculus. But the fact is, despite the sensitivity to initial conditions, there is a pattern generated by chaotic orbits. As we continue our investigations in this series of books, we will find many other examples of patterns within chaos, including the orbit diagram, the Julia sets, and the Mandelbrot set.

NAME(S):

1 ▹ SENSITIVITY TO INITIAL CONDITIONS

Compute the orbits of the neighboring seeds 0, 0.01, 0.00001, and 0.000001 under the logistic iteration rule $x \rightarrow 4x(1 - x)$. How many iterations does it take for corresponding points on these orbits to differ by 0.5?

____

____

2 ▹ GRAPHICAL ITERATION AND SENSITIVITY

Graphical iteration can be used to illustrate sensitivity to initial conditions. Here is the graph of $y = 4x(1 - x)$ with a particular seed, x_0, displayed. Using a ruler, carefully sketch the first 10 points on the orbit of x_0. Remember that you should always go vertically to the graph first, then horizontally back to the diagonal. We have superimposed a grid on this figure to help you with the sketch.

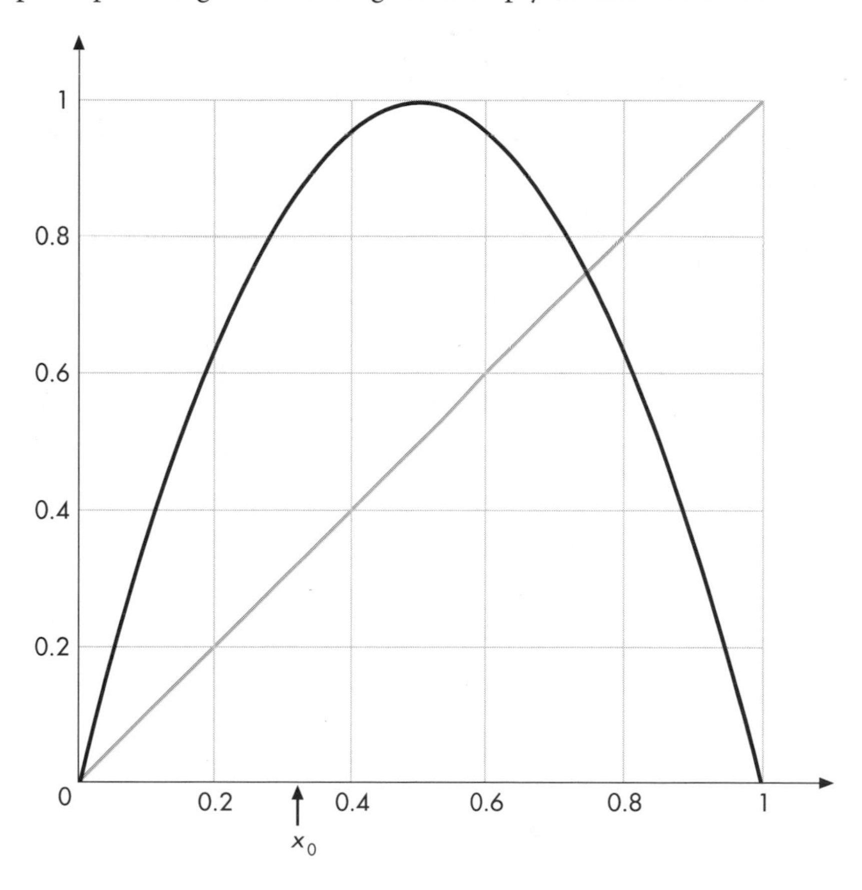

After 12 iterations, mark the point on the x-axis your orbit has reached. That is, mark the value of x_{12} on the x-axis. Now put this sheet away and perform the same experiment with this next graph (same seed), or compare your results with those of your classmates. Are your x_{12}-values close together? Explain what you think is happening.

____

____

____

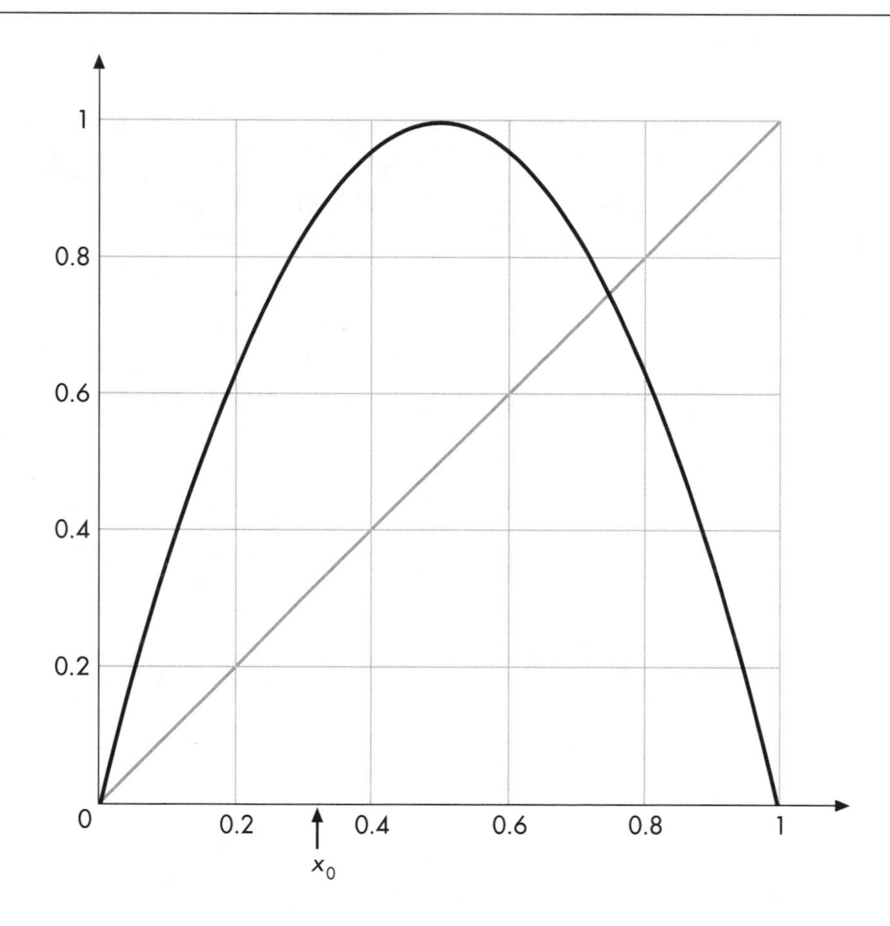

3 ▹ ANOTHER LOGISTIC ITERATION RULE

Consider the logistic iteration rule $x \to 3.9x(1 - x)$.

Choose any two seeds within the interval $0 < x_0 < 1$ that differ from each other by only 0.00001. Then compute the orbits of both seeds. Do they eventually differ from one another?

At which iteration do they differ from one another by 0.1?

Do they ever differ by 0.5? If so, at which iteration?

Plot the time-series graph for each of your orbits in different colors, indicating the iteration at which the two orbits differ by 0.1 and 0.5.

4 ▹ MORE LOGISTIC ITERATION RULES

Which of the following logistic iteration rules depend sensitively on initial conditions? In each case, select several nearby seeds in the interval $0 < x < 1$ and compute the first 50 iterations before making your decision.

a. $x \to 3.2x(1 - x)$

b. $x \to 3.83x(1 - x)$

c. $x \to 3.7x(1 - x)$

d. $x \to 3.93x(1 - x)$

5 ▹ A PROJECT

If you and your classmates have access to several different kinds of computers and/or calculators, use each kind to compute the orbit of the seed 0.123 under the logistic iteration rule $x \rightarrow 4x(1 - x)$. Are these orbits the same or (eventually) different?

6 ▹ SENSITIVITY IN THE REAL WORLD

Does a pendulum exhibit sensitivity to initial conditions? Obviously not. But what happens if you drop a pencil point-first toward the ground? Does the final resting place change if you drop the pencil again? What happens if you hold a piece of paper horizontally and drop it several times from the same height? Find other examples of physical systems, some that are sensitive and some that are insensitive to initial conditions.

1. Which of the following iteration rules exhibit sensitivity to initial conditions for seeds chosen in the interval $-1 < x < 1$?

   a. $x \rightarrow x^2 - 2$

   b. $x \rightarrow x^2 - 1$

   c. $x \rightarrow x^2 - 1.5$

   d. $x \rightarrow x^2 - 1.3$

2. Consider the iteration rule (called the **doubling rule mod 1**)

$$x \rightarrow \begin{cases} 2x & \text{if } 0 < x < \frac{1}{2} \\ 2x - 1 & \text{if } \frac{1}{2} \leq x < 1 \end{cases}$$

   Does this iteration rule exhibit sensitivity to initial conditions? Answer this by first using graphical iteration and then using a computer to list orbits. Do your results agree?

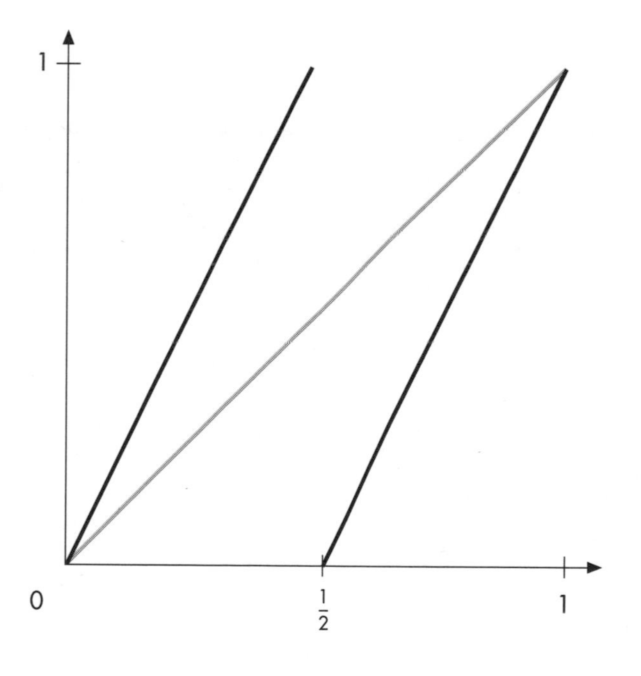

TEACHER NOTES

Cycles and Nonlinear Iteration

OVERVIEW

In this lesson, we move beyond fixed points and try to understand the behavior of cycles in nonlinear iteration rules. This is a fairly robust lesson that is not intended for the mathematically faint at heart! Much of the material is optional, and we provide a modified outline of a trimmed-down discussion of this topic below.

MATHEMATICAL PREREQUISITES

This lesson interweaves virtually all of the material from previous lessons: computing orbits, viewing orbits as time-series graphs and web diagrams, plotting graphs of higher iterations, and engaging in some fairly sophisticated algebra.

MATHEMATICAL CONNECTIONS

Students need to have a good grasp of **composition of functions** to understand the meaning of the graphs of higher iterations. Also, the algebra involved in finding 2-cycles is very challenging. It involves **factoring, division of polynomials**, and use of the **quadratic formula** involving parameters in place of actual numbers.

TECHNOLOGY

This lesson requires the use of some type of technology to perform the Investigations. A graphing calculator is sufficient for the graphical Investigations. A computer algebra system can be used to effect some of the difficult algebraic steps. The nonlinear Web applet at our Web site allows students to see the graphs of the first several iterations of certain rules. And the target and cycle practice applets allow them to search for these cycles interactively.

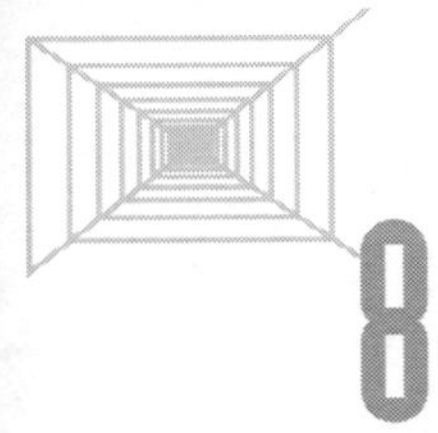

SUGGESTED LESSON PLAN

CLASS TIME

Two 50-minute class periods for the full chapter, or one 50-minute session for the pared-down version.

PREPARATION

The main goal is for students to see that the logistic iteration rule may possess cycles of various periods. This can be accomplished by doing Investigations 1–3 either in class or as homework. The students will see that different *k*-values lead to cycles of different periods for the logistic iteration rule. Covering this material allows the student to move ahead to the next lesson, where the orbit diagram is discussed.

Most of the chapter is devoted to why the logistic iteration rule leads to cycles. Demonstrating that involves either plotting the graphs of higher iterations or doing the necessary algebra (or both!). We suggest that you use the transparency masters provided, superimposing the graph of the second iteration over the original graph, to help students see the relationship.

LESSON DEVELOPMENT

Begin by having students perform Investigations 1–3 and summarizing their results. Investigation 2 in particular leads students through the birth of a 2-cycle (the period-doubling bifurcation).

This can be illustrated by drawing the graphs of the second iterations of each of the rules involved, as in Investigation 4. Students have a difficult time understanding the relationship between the graph of the first iteration and that of the second. It is worthwhile pointing out places where both graphs have fixed points (the original fixed points) as well as places where the second iteration has fixed points but the first iteration does not. These are the 2-cycles. Superimposing these two graphs, as in Transparencies 8A and 8B, helps.

We can also find the formula for the second iteration explicitly. Students may find the algebra difficult in that it involves

1. knowing that fixed points are also fixed points for the second iteration,
2. computing the formula of the second iteration, and
3. using long division to eliminate the fixed points from the expression for the second iteration.

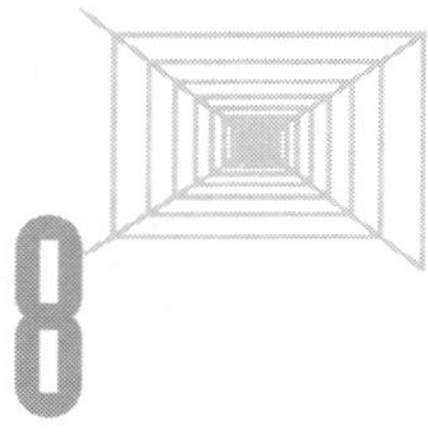

Students who can handle all of this have an excellent grasp of the meaning of iteration. In combination with the graphical approach above, this algebra provides a wonderful link between the geometric and algebraic approach to iteration.

The target practice in Investigations 6 and 7 provides students with a fun way of searching for cycles. You may use the applets "Target Practice" and "Cycle Practice" on our Web site **http://math.bu.edu/DYSYS/applets/** to do this interactively.

LESSON NOTES

The section on algebraically finding the 2-cycles can safely be skipped. Even if this material is covered, be sure to explain that this method fails for finding higher period cycles. Indeed, just to find 3-cycles, we need to solve a polynomial equation of degree 8 (though we can lower it to 6 by dividing out the fixed points). Still, no help!

Students seem to get a lot out of plotting the third iteration of the logistic rule for values of k in the vicinity of 3.83 to see the emergence of a 3-cycle as discussed in the Explanation. This will relate to the opening of the period 3 window in the orbit diagram, as discussed in the next lesson.

Further Exploration problem 4, dealing with the doubling function, also provides a nice way to see the interplay between algebra and geometry. In this case, you can explicitly write down all the cycles of period n and, moreover, understand the graph of the nth iteration completely.

The graphs of the first and second iterations of the rule $x \to 3.2x(1 - x)$

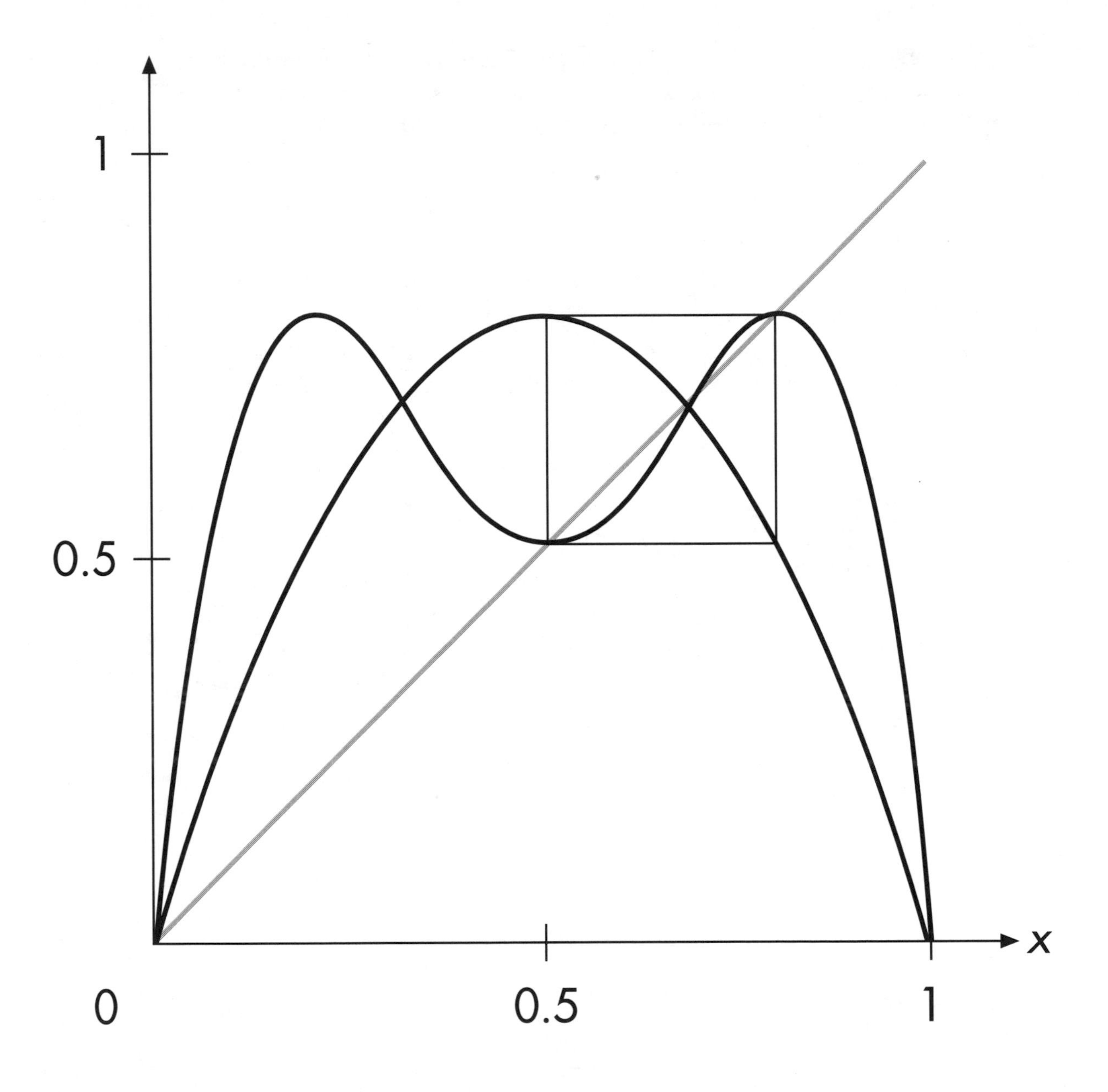

The graphs of the first and second iterations of the rule $x \to 2.8x(1 - x)$

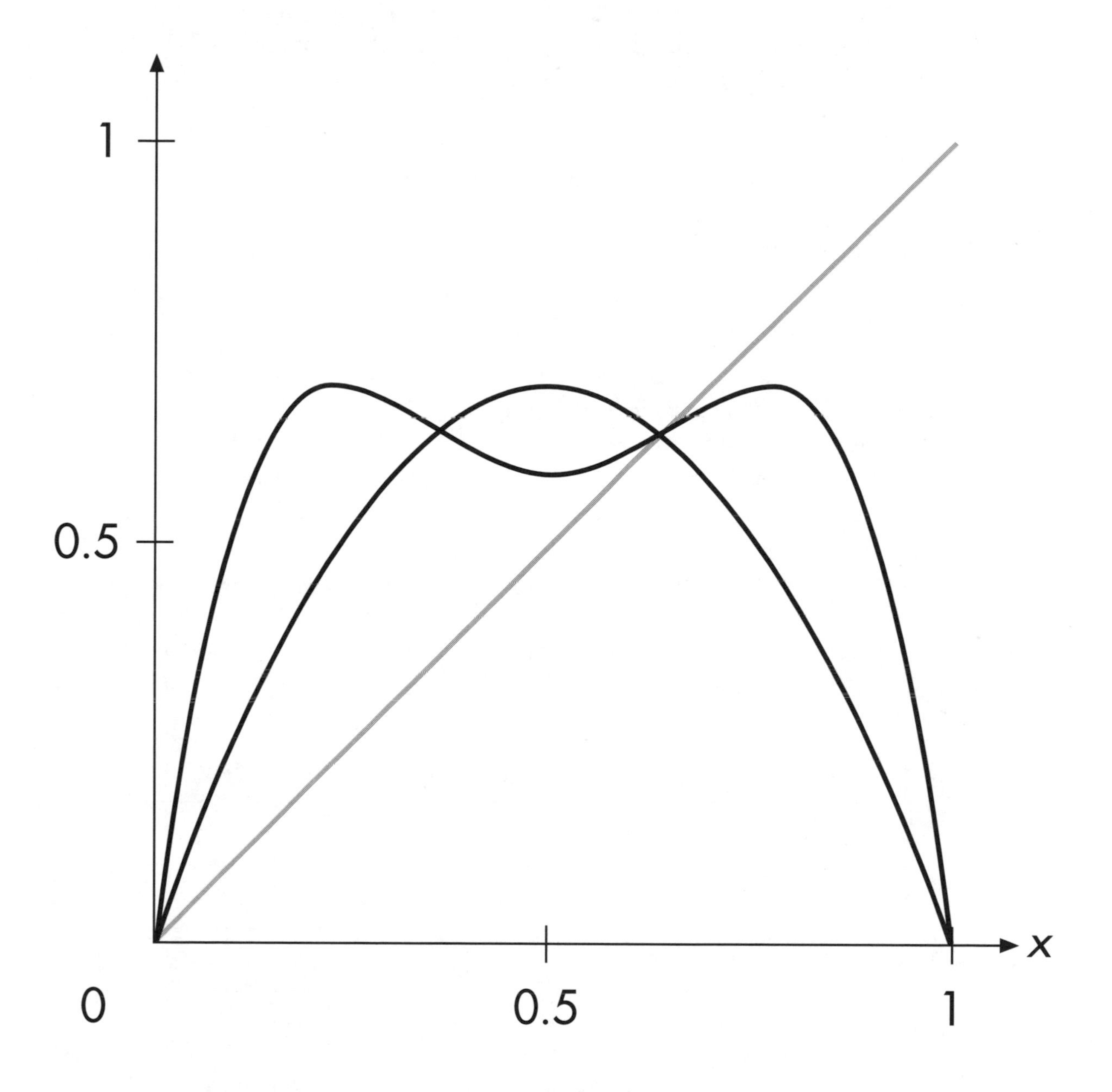

As we have seen, there are many different types of orbits for nonlinear iteration rules. In the last few lessons, we examined the simplest types of orbits, the fixed points, as well as very complicated orbits that depend quite sensitively on initial conditions. In this lesson, we discuss cycles, another important type of orbit. Recall that a **cycle** is an orbit that repeats, like the orbit of 0 for the iteration rule $x \to x^2 - 1$, which is

$$0 \to -1 \to 0 \to -1 \to \cdots$$

This is a **cycle of period 2**, or a **2-cycle**, since the orbit repeats at every second iteration. As we will see below, iteration rules may have many cycles with many different periods.

THE LOGISTIC ITERATION RULE

In previous lessons, we saw that there was a significant difference between the fate of orbits for the logistic iteration rule $x \to 2.8x(1 - x)$ and the fate of the orbits for $x \to 3.2x(1 - x)$. Here is a table of the first 30 points (to 3 decimal places) on the orbit of the seed $x_0 = 0.12$ for each of these iteration rules:

	$k = 2.8$	$k = 3.2$
	0.12	0.12
1	0.296	0.338
2	0.583	0.716
3	0.681	0.651
4	0.609	0.727
5	0.667	0.635
6	0.622	0.742
7	0.658	0.613
8	0.630	0.759
9	0.653	0.585
10	0.635	0.777

	$k = 2.8$	$k = 3.2$
11	0.649	0.555
12	0.638	0.791
13	0.647	0.530
14	0.640	0.797
15	0.646	0.518
16	0.641	0.799
17	0.645	0.514
18	0.641	0.799
19	0.644	0.513
20	0.642	0.799
21	0.644	0.513

	$k = 2.8$	$k = 3.2$
22	0.642	0.799
23	0.643	0.513
24	0.642	0.799
25	0.643	0.513
26	0.643	0.799
27	0.643	0.513
28	0.643	0.799
29	0.643	0.513
30	0.643	0.799

Notice that the orbit tends to a fixed point when $k = 2.8$, but the orbit tends to a 2-cycle when $k = 3.2$.

Graphical iteration provides a different way to view this: When $k = 2.8$, the orbit tends to an attracting fixed point.

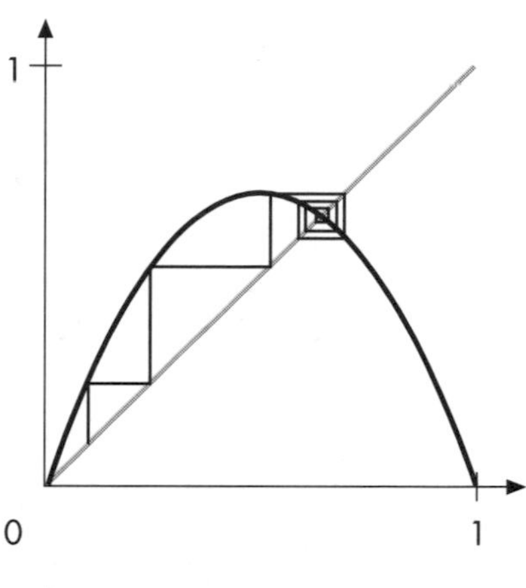

But when $k = 3.2$, the orbit tends to a 2-cycle:

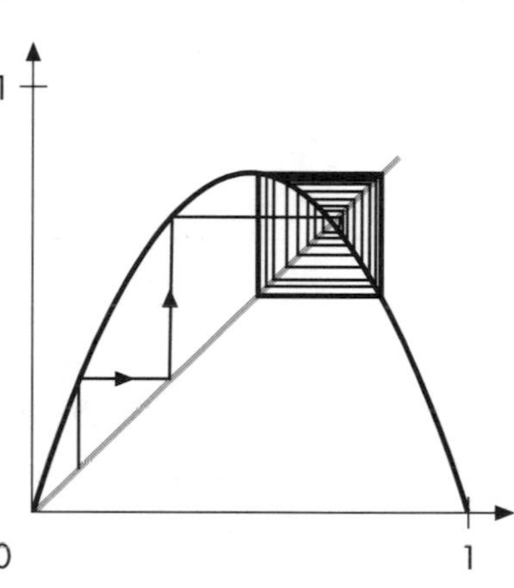

If you follow this orbit, you see that it first comes close to the (repelling) fixed point. Then it moves slowly away until it begins to approach a 2-cycle. This cycle is easier to see if we remove the transient behavior. If we plot only iterations 25 to 30 on this orbit, we see the 2-cycle displayed as a box:

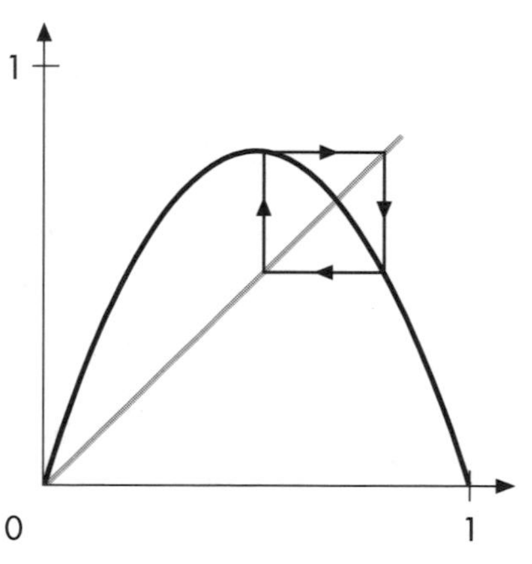

As a time series, the full orbit looks like this:

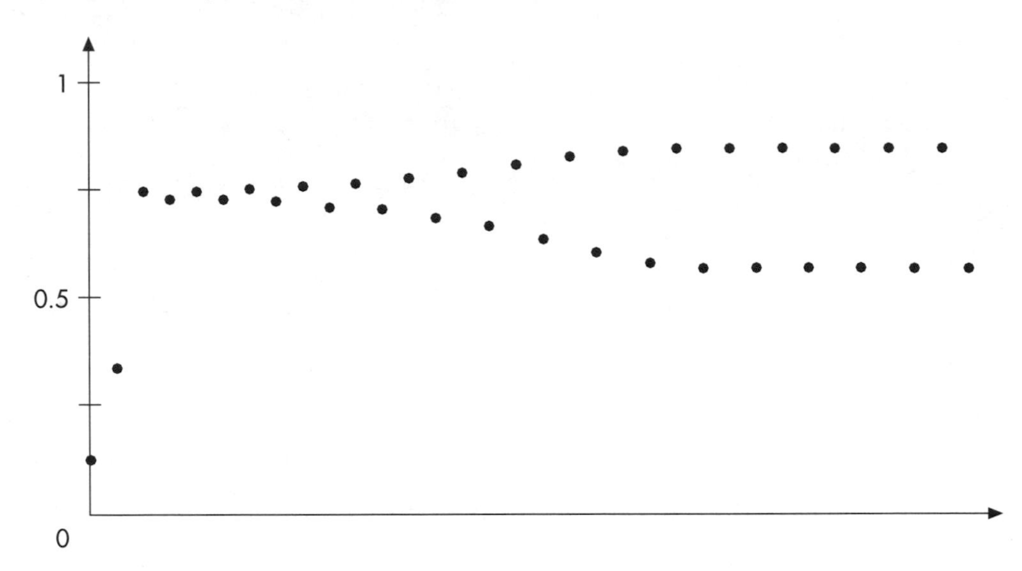

FINDING 2-CYCLES

Let's try to determine where the 2-cycle for the logistic iteration rule comes from. We have several options as to how to proceed. Just as fixed points may be located algebraically by solving the equation $kx(1 - x) = x$, we can also find period 2 points by finding those x-values for which the second iteration has a fixed point. This means we must find the places where $x_2 = x_0$ so that the orbit becomes

$$x_0 \to x_1 \to x_0 \to x_1 \to \cdots$$

That is, we must find the places where the graph of the second iteration of the logistic iteration rule crosses the diagonal. So we must first compute the second point on the orbit of a typical seed x_0. That's not so bad. We compute the first iteration

$$x_1 = kx_0\left(1 - x_0\right)$$

and then the second

$$x_2 = kx_1\left(1 - x_1\right) = k\left[kx_0\left(1 - x_0\right)\right]\left[1 - kx_0\left(1 - x_0\right)\right]$$

That is, the iteration rule for the second iteration is

$$x \to k[kx(1 - x)][1 - kx(1 - x)]$$

We must therefore find the places where the graph of

$$y = k[kx(1 - x)][1 - kx(1 - x)]$$

meets the diagonal line $y = x$. That is, we must solve the equation

$$k[kx(1 - x)][1 - kx(1 - x)] = x$$

This does not look like an easy equation to solve. If you multiply out the left side, you will find a fourth-degree polynomial. So let's put that algebra on hold for a while and attack this problem in a different way.

Graphing the Second Iteration

The graph of $y = k[kx(1 - x)][1 - kx(1 - x)]$ is easy to plot using a graphing calculator. Let's plot this graph when $k = 2.8$ as well as the graph of the first iteration $y = 2.8x(1 - x)$.

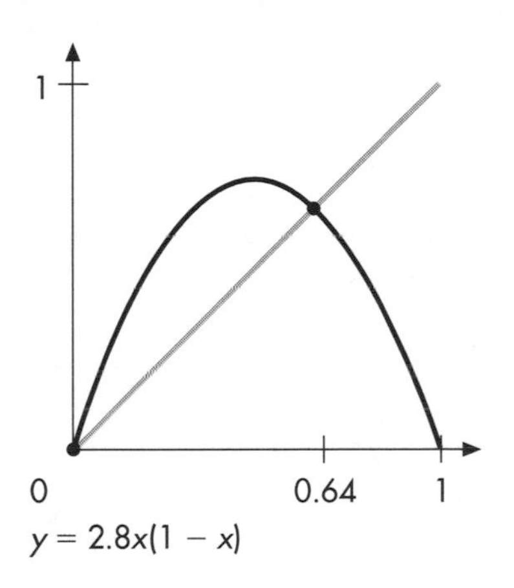

$y = 2.8x(1 - x)$

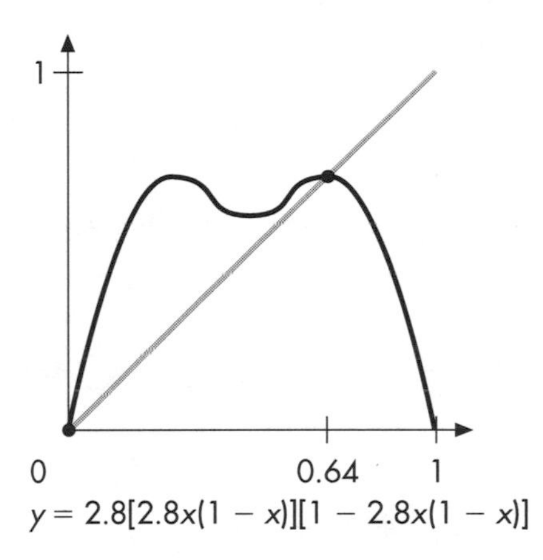

$y = 2.8[2.8x(1 - x)][1 - 2.8x(1 - x)]$

Now you can compute that the iteration rule $x \rightarrow 2.8x(1 - x)$ has a fixed point at 0 and at $^{18}/_{28}$, or 0.64.

Note that each of these graphs crosses the diagonal at these points. So the points of intersection of the second iteration with the diagonal are actually just the fixed points for the first iteration. This changes, however, when we plot the corresponding graphs for $k = 3.2$. Now the nonzero fixed point for the iteration rule $x \rightarrow 3.2x(1 - x)$ is located at $^{2.2}/_{3.2} = 0.6875$.

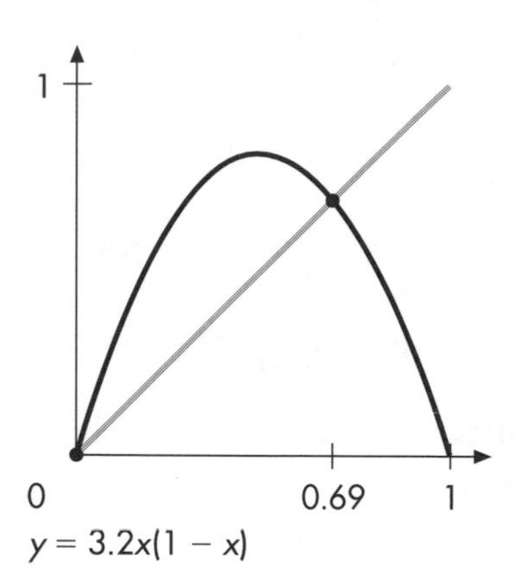

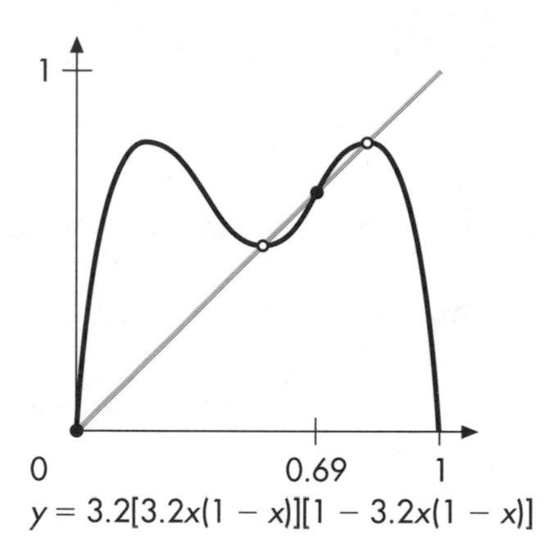

Now, however, there are two new places where the graph of the second iteration crosses the diagonal. These cannot be fixed points for $x \rightarrow 3.2x(1 - x)$. Therefore, they must be points that lie on a 2-cycle.

If we superimpose both of these graphs, we see how the points on the 2-cycle relate to the fixed points given by graphical iteration of $x \rightarrow 3.2x(1 - x)$:

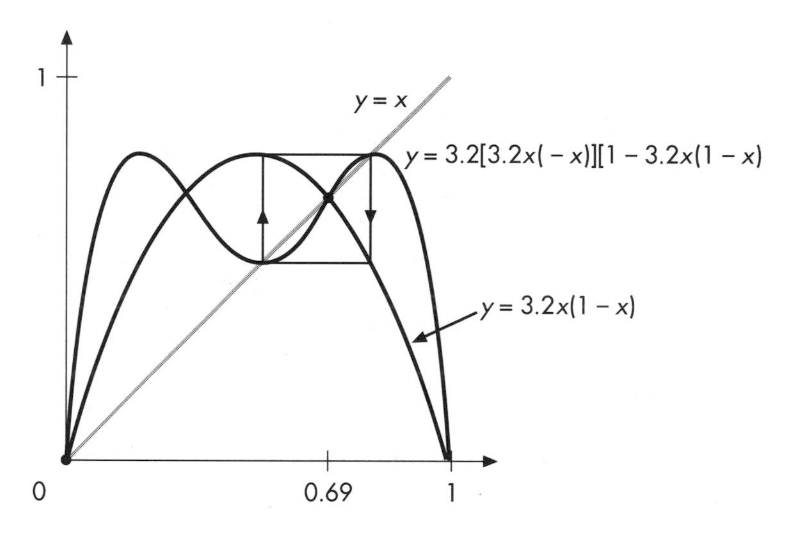

FINDING CYCLES GEOMETRICALLY

Plotting the graphs of the higher iterations of the logistic iteration rule is an easy way to see if the iteration rule has cycles. To simplify terminology, let's use function notation to name these graphs. Say the first iteration is given by

$$F(x) = kx(1 - x)$$

Then the second iteration is given by $F(F(x)) = F(kx(1 - x))$. That is,

$$F(F(x)) = k[kx(1 - x)][1 - kx(1 - x)]$$

We can continue in this fashion and compute the expressions for higher iterations. For example,

$$\begin{aligned} F(F(F(x))) &= F(k[kx(1 - x)][1 - kx(1 - x)]) \\ &= k \cdot (k[kx(1 - x)][1 - kx(1 - x)]) \cdot [1 - (k[kx(1 - x)][1 - kx(1 - x)])] \end{aligned}$$

This is not an easy formula to work with or to graph; however, there are many software packages and graphing calculator routines that allow you to draw the graph of higher iterations of functions or iteration rules. For example, you can use the applet "Nonlinear Web" at our Web site **http://math.bu.edu/applets/** to plot higher iterations of many of the iteration rules discussed in this book.

Here is the plot of $F(F(F(x)))$ when $k = 3.7$. Note that this graph crosses the diagonal in only two places: at 0 and at one other point, which must be the second fixed point for this iteration rule. We can therefore conclude that the iteration rule $x \rightarrow 3.7x(1 - x)$ has no 3-cycles in the interval $0 \leq x \leq 1$.

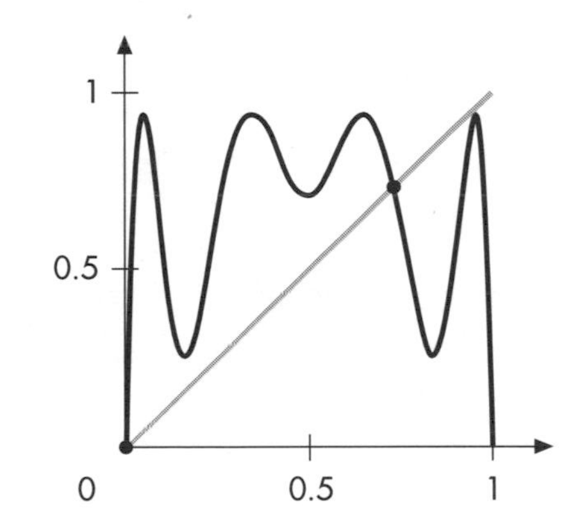

On the other hand, the graph of the third iteration of the iteration rule $x \rightarrow 3.9x(1 - x)$ appears to cross the diagonal many more times (eight, in fact). Two of these points are the original fixed points; the other six points must lie on a 3-cycle, as they cannot lie on a 2-cycle and also be fixed under the third iteration.

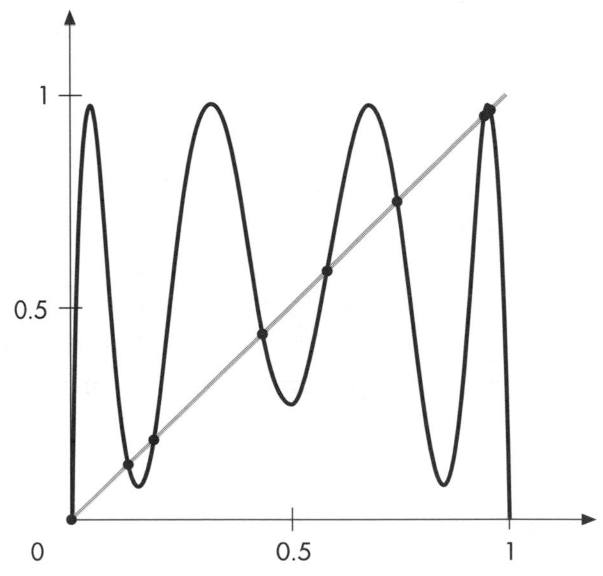

THE CHAOTIC LOGISTIC ITERATION RULE

When we iterated the logistic iteration rule $x \rightarrow 4x(1 - x)$, we rarely found any cycles with the exception of the fixed points at 0 and $^3/_4$. However, they are there, as graphical methods allow us to see. Here are the graphs of the first, second, and third iterations of this rule:

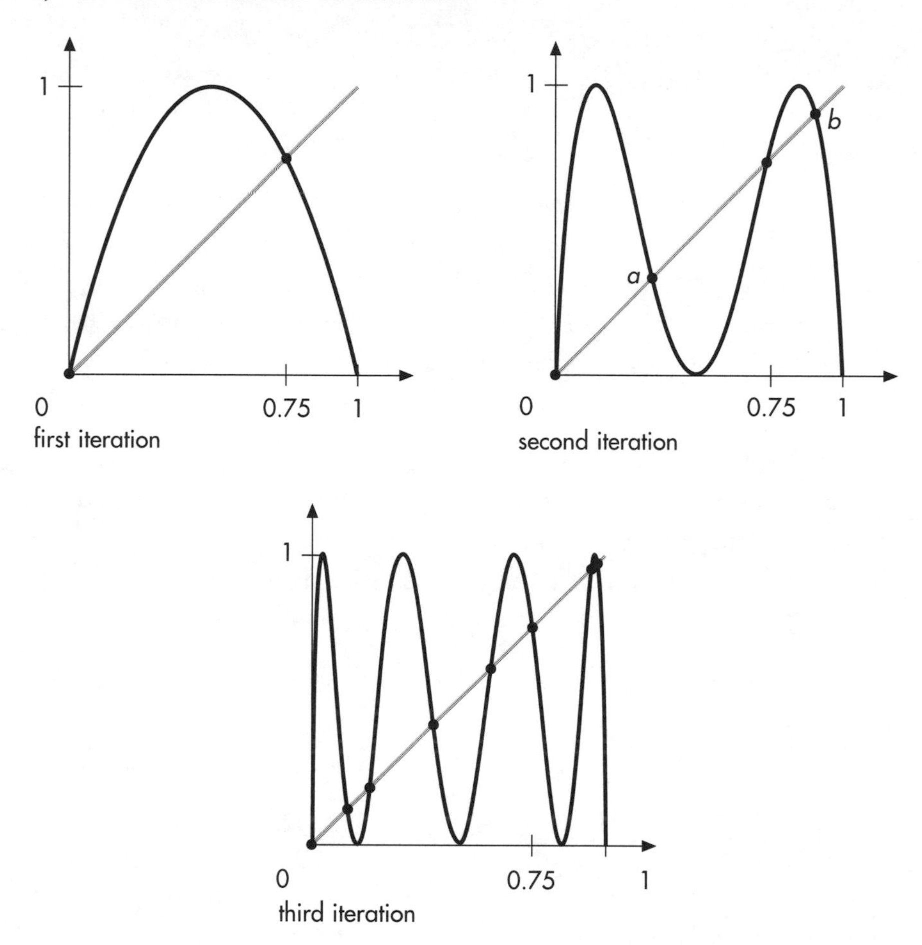

The points marked a and b on the graph of the second iteration are not fixed points, so they must lie on a cycle of period 2. Similarly, all six points on the graph of the third iteration that are not fixed points cannot lie on 2-cycles (a 2-cycle is not fixed under the third iteration!), so these points must lie on 3-cycles. Notice that for this iteration rule we have two points on a 2-cycle and six points on a 3-cycle.

TARGET PRACTICE

As we have seen, the chaotic logistic iteration rule $x \rightarrow 4x(1 - x)$ must have cycles. As we will see at right, finding cycles algebraically is next to impossible, but with a little geometry and a good eye, we can find them (at least approximately) using graphical iteration. To do this, we set up the graphical iteration and use trial and error to find points on an n-cycle. For example, to find a 2-cycle, we apply graphical iteration twice and try to end up where we started. If we begin with the seed $x_0 = 0.3$, we see that our guess is off by a bit (the second iteration is too large):

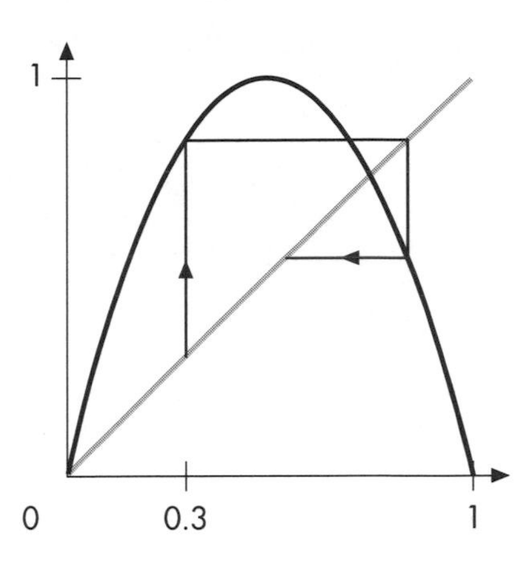

If instead we choose $x_0 = 0.4$, the second iteration is now too small:

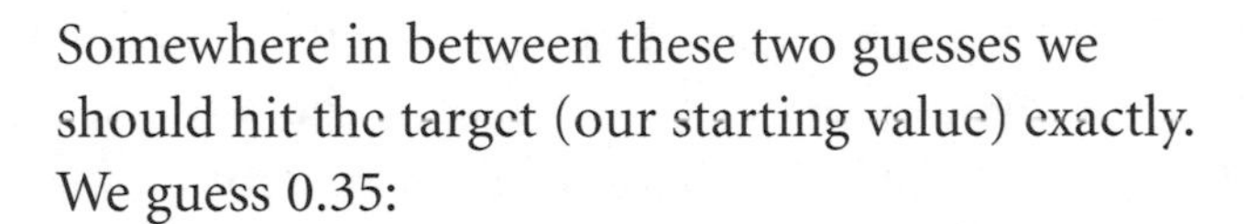
Somewhere in between these two guesses we should hit the target (our starting value) exactly. We guess 0.35:

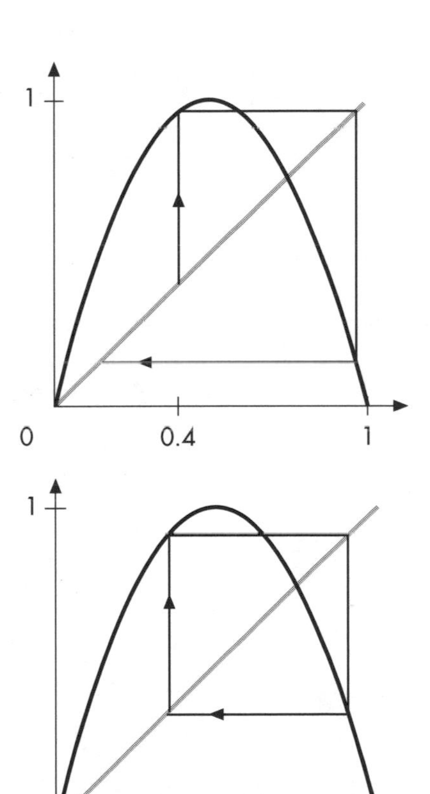

Very close, but still a little too small.
So we try 0.34:

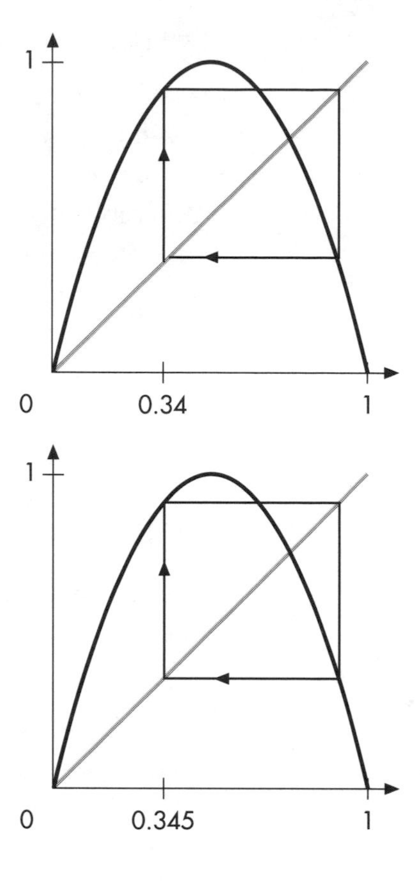

A little too high. Finally, the seed 0.345:

It is just about right. We have found one point on a 2-cycle for the iteration rule $x \to 4x(1 - x)$. (The actual value is 0.34549 . . . .)

NOW FOR THE ALGEBRA

After all of this geometry, let's come back and determine some of these cycles algebraically. Let's try to determine exactly when the logistic iteration rule $x \to kx(1 - x)$ has a 2-cycle. The idea behind this is exactly the same as finding fixed points. For fixed points, we asked for x-values where the graph of the first iterate of our rule met the diagonal. That is, we had to solve the equation $kx(1 - x) = x$. This is a quadratic equation, so solving it is easy. In Lesson 5 we determined that the two roots of this equation are 0 and $\frac{k-1}{k}$. So the fixed points are 0 and $\frac{k-1}{k}$.

To find the 2-cycles, we must find the places where the graph of the second iteration of this rule crosses the diagonal. As we have seen, this means we must find the places where the graph of

$$y = k[kx(1 - x)][1 - kx(1 - x)]$$

meets the diagonal line $y = x$. That is, we must solve the equation

$$k[kx(1 - x)][1 - kx(1 - x)] = x$$

Now let's do this algebra. Hold on to your hats; the going will get a bit rough here. Multiplying out the terms on the left side, we find

$$k\left(kx - kx^2\right)\left(1 - kx + kx^2\right) = x$$

or

$$k\left(kx - k^2x^2 + k^2x^3 - kx^2 + k^2x^3 - k^2x^4\right) = x$$

Continuing on fearlessly, we see that this simplifies (somewhat) to

$$\left(k^2 - 1\right)x - \left(k^3 + k^2\right)x^2 + 2k^3x^3 - k^3x^4 = 0$$

Holy mackerel! This is a fourth-order equation whose roots we must find. Worse than that, the coefficients are not given numbers. Rather, each of them involves the parameter k. This looks bad. But wait a minute. We see right away that we can find one root of this equation: One solution is clearly $x = 0$. So we can factor out x from this equation and find the simpler (third-degree) equation:

$$\left(k^2 - 1\right) - \left(k^3 + k^2\right)x + 2k^3x^2 - k^3x^3 = 0$$

The reason 0 is a solution of this equation is that 0 is a fixed point for the iteration rule $x \rightarrow kx(1 - x)$. As such, it is also fixed by the second iteration. Aha! That gives us another root for the fourth-order equation, for we know that $^{k-1}/_k$ is also a fixed point. Therefore, $^{k-1}/_k$ also satisfies this equation. You can check this by inserting $x = {}^{k-1}/_k$ into the preceding equation and solving.

So we know that the equation

$$\left(k^2 - 1\right) - \left(k^3 + k^2\right)x + 2k^3x^2 - k^3x^3 = 0$$

can be factored into

$$\left(x - \frac{k-1}{k}\right) \cdot \text{(something)}$$

and the "something" is going to be a quadratic expression. So we win! All we have to do is find the "something" and set it equal to 0. The roots of this (much simpler) equation will give us our 2-cycle (at last).

How do we find the "something"? Well, we could use long division. That will work, but it looks messy. On the other hand, we know that the "something" is a quadratic expression and so must be of the form $Ax^2 + Bx + C$. We need to find A, B, and C so that

$$\left(x - \frac{k-1}{k}\right) \cdot \left(Ax^2 + Bx + C\right) = \left(k^2 - 1\right) - \left(k^3 + k^2\right)x + 2k^3x^2 - k^3x^3$$

Multiplying out the left side yields

$$Ax^3 + \left(B - \frac{k-1}{k}A\right)x^2 + \left(C - \left(\frac{k-1}{k}\right)B\right)x - \left(\frac{k-1}{k}\right)C =$$

$$-k^3x^3 + 2k^3x^2 - \left(k^3 + k^2\right)x + \left(k^2 - 1\right)$$

Collecting like terms on the left, we first see that the cubic terms are Ax^3 and $-k^3x^3$ and hence $Ax^3 = -k^3x^3$. So we must have $A = -k^3$.

Setting the quadratic terms equal to each other, we must have

$$\left(B - \frac{k-1}{k}A\right) = 2k^3$$

Substituting $A = -k^3$ into this expression and solving, we find that $B = k^3 + k^2$.

Finally, collecting the constant terms, we find

$$-\frac{k-1}{k}C = k^2 - 1$$

Dividing by $(k - 1)$ and simplifying, we then find $C = -k^2 - k$.

We've done it! This means that the "something" expression is

$$-k^3x^2 + \left(k^3 + k^2\right)x - \left(k^2 + k\right)$$

We must set this expression equal to 0 and solve. Dividing first by k, we must solve the quadratic equation

$$-k^2x^2 + \left(k^2 + k\right)x - (k + 1) = 0$$

That's now a piece of cake because we have the quadratic formula at our disposal. Using the quadratic formula, we find that the roots are

$$-(k^2 + k) \pm \frac{\sqrt{(k^2 + k)^2 - 4k^2(k + 1)}}{-2k^2}$$

Simplification yields

$$\frac{-(k^2 + k) \pm \sqrt{k^4 - 2k^3 - 3k^2}}{-2k^2}$$

or

$$\frac{-(k^2 + k) \pm k\sqrt{(k + 1)(k - 3)}}{-2k^2}$$

In order for this expression to give real roots, the product inside the radical must be greater than or equal to 0. So we must have either $k \geq 3$ or $k \leq -1$. Since we are dealing only with the case of k positive, it follows that this quadratic equation has no roots when $0 \leq k < 3$, one root when $k = 3$, and two roots when $k > 3$. In this latter case, the expression above gives us the two roots. Since these roots are not fixed points, they must be our 2-cycle. So we are done. Whew!

THE MORAL OF THE STORY

That was indeed a difficult equation to solve. It has called into play almost all of our algebraic skills. And yet, all we found were the fixed points and 2-cycles. Imagine now what would happen if we tried to find the 3-cycles algebraically. We won't even bother to write down the equation for the third iteration of the rule $x \rightarrow kx(1 - x)$, but should you try to write it down, you will find that it is a polynomial equation of degree 8. Just as before, two of the roots are the fixed points, which we already know. Unfortunately, the 2-cycles will not be roots (they are not fixed by the third iteration). So we are left with an intimidating sixth-degree equation that is impossible to solve.

Here we see an important point in studying iteration. We can find cycles (at least approximately) using geometry (either the graphs of higher iterations or our target method) or using the computer (numerical experimentation). However, algebra does not help as much as we would like. The moral of the story is: When algebra works and works easily, use it. But when the algebra gets messy or impossible, remember that you have other tools available.

NAME(S):

1 ▹ FINDING CYCLES

For each of the following iteration rules, compute the first 25 points on the orbit of 0.5 and then display this orbit using graphical iteration. Does the orbit tend to a cycle? If so, show the fate of the orbit by removing the transient behavior.

a. $x \rightarrow 3.3x(1 - x)$ ______

b. $x \rightarrow 3.5x(1 - x)$ ______

c. $x \rightarrow 3.55x(1 - x)$ ______

d. $x \rightarrow 3.83x(1 - x)$ ______

2 ▹ FIXED POINTS AND 2-CYCLES

Let's look in more detail at the iteration rule $x \rightarrow kx(1 - x)$ when k lies between 2.8 and 3.2. Using a computer or calculator, compute the fate of the orbit of the seed $x_0 = 0.5$ for at least 10 different k-values in this interval. Record the k-values for which this orbit tends to a fixed point as well as those for which it tends to a 2-cycle. Also record the k-values for which this information is too difficult to discern.

Tends to fixed point: ______

Tends to 2-cycle: ______

Cannot determine: ______

3 ▹ FINDING OTHER CYCLES

Consider the same question as in Investigation 2, but this time using k-values from the interval 3.4 to 3.5. You should basically see two different fates of the orbit of 0.5, and a collection of k-values in between where the fate of this orbit is not clear. Describe your findings briefly:

______

______

______

4 ▹ FINDING CYCLES GRAPHICALLY

Plot the graphs of $y = kx(1 - x)$ and the second iteration of this rule for a variety of k-values between 2.8 and 3.2. Discuss the behavior of these graphs. For which (approximate) k-value do you begin to see a change in the number of intersections with the diagonal? Explain what is happening as k increases through this value.

5 ▹ HIGHER ITERATIONS

a. Plot the graphs of the fourth and fifth iterations of the iteration rule $x \to 4x(1 - x)$. How many times do these graphs cross the diagonal?

b. Using this information, what do you expect the graph of the nth iteration of this rule to look like? How many times do you think this graph will cross the diagonal?

6 ▹ FINDING A 3-CYCLE

Find the approximate location of a 3-cycle for the iteration rule $x \to 4x(1 - x)$ using the "target practice method."

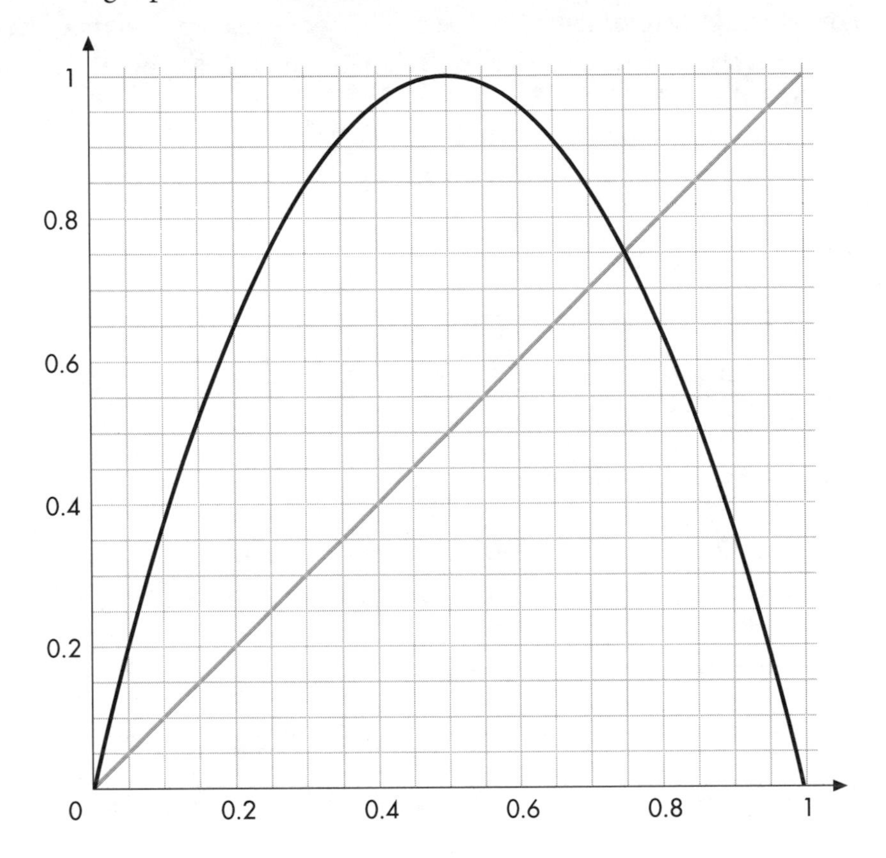

You may also use the target practice applet located on the Web site at **http://math.bu.edu/DYSYS/applets**.

7 ▹ Finding a 4-cycle

Find the (approximate) location of a 4-cycle for the iteration rule $x \rightarrow 4x(1 - x)$ using the target practice method as in Investigation 6. Using the results of Investigation 6, how many distinct points that have period 4 do you expect to find?

____

____

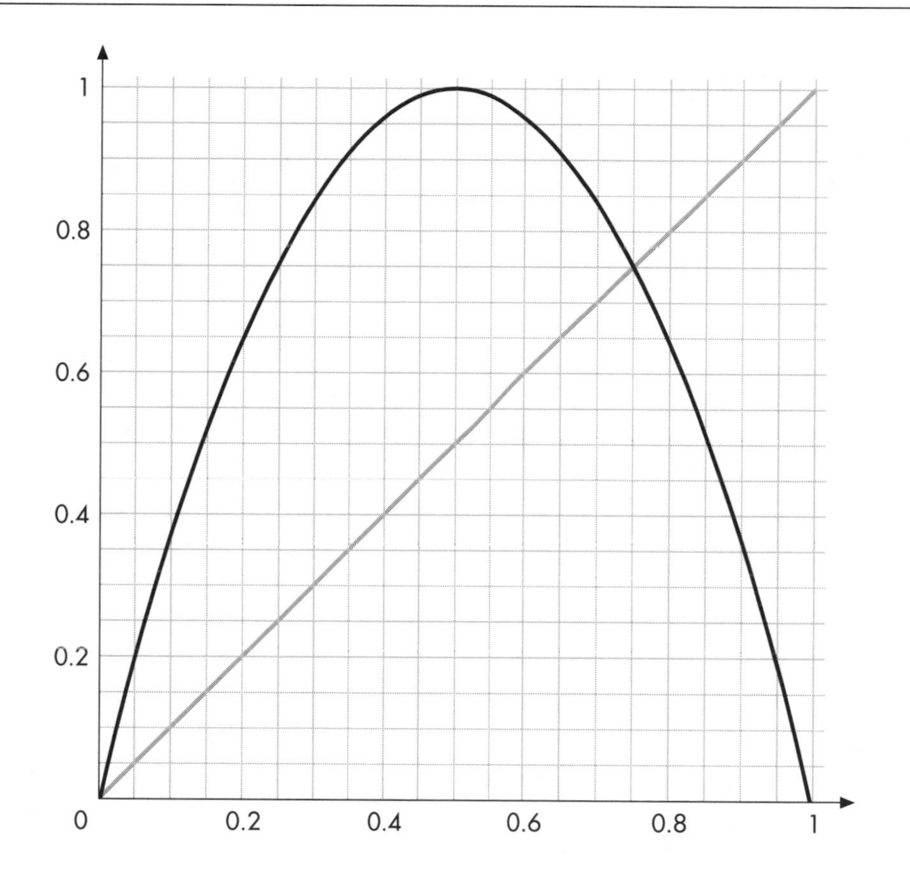

1. Draw the graph of the first iteration of the rule $x \to x^2 - 2$ over the interval $-2 \le x \le 2$. Then draw the graphs of the second and third iterations of this rule over the same interval. How many times does each iteration cross the diagonal? What are the periods of these cycles?

2. Find algebraically the points of period 2 for the iteration rule $x \to x^2 - 2$.

3. Use graphical iteration to explain the fate of all orbits of the iteration rule $x \to -x^3$. What is the 2-cycle for this rule?

4. Consider the iteration rule (called the **doubling rule mod 1**)

$$x \to \begin{cases} 2x & \text{if } 0 < x < \frac{1}{2} \\ 2x - 1 & \text{if } \frac{1}{2} \le x < 1 \end{cases}$$

   a. Use our target practice method to locate a cycle of period 3 for this iteration rule. Can you find this cycle explicitly?

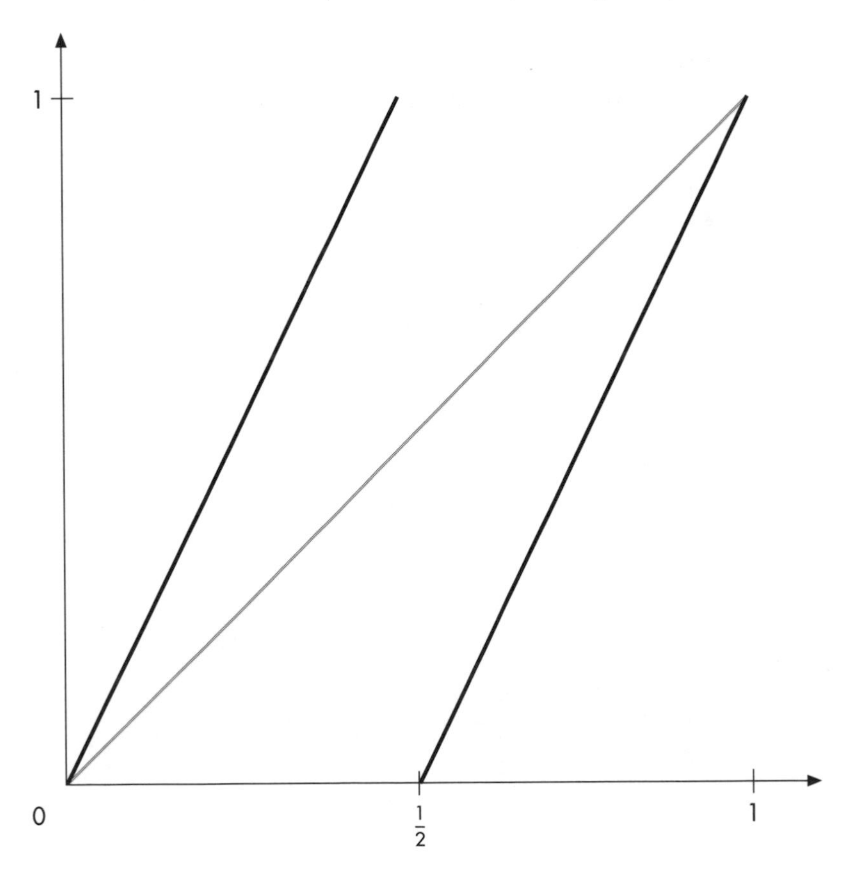

b. Now use the target practice method to locate a cycle of period 4 for this iteration rule.

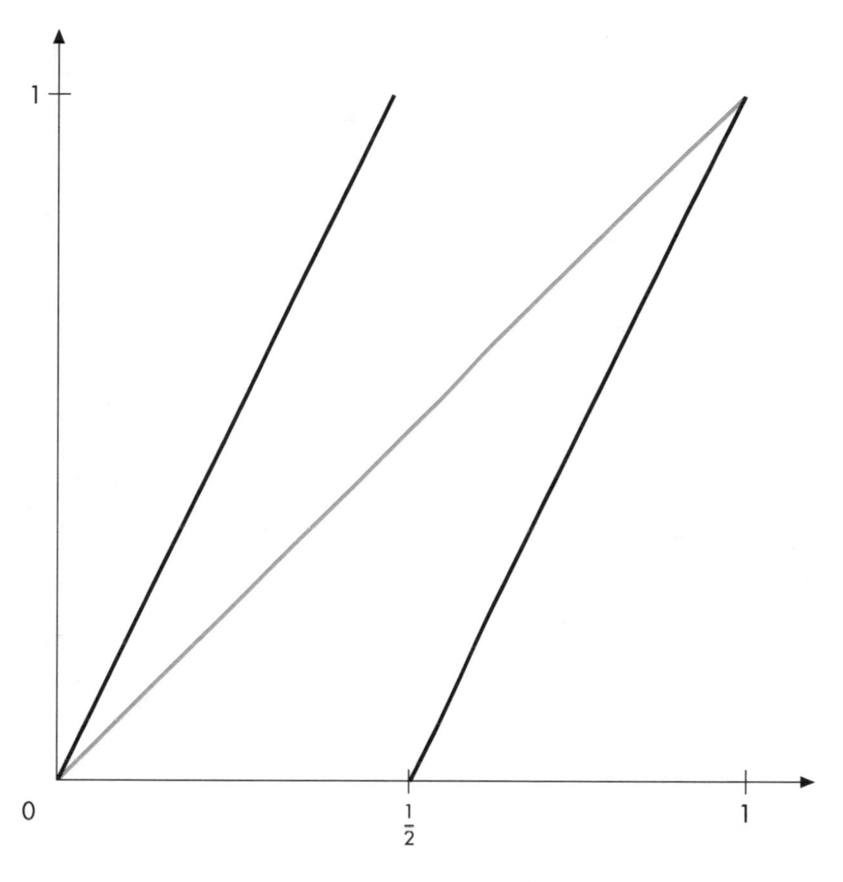

c. Sketch the graphs of the second and third iterations of the doubling rule. How many times does each graph cross the diagonal?

TEACHER NOTES

The Orbit Diagram

Overview

The goal of this lesson is to put most of the information in the previous lessons together to view the orbit diagram for the logistic iteration rule. This picture is a summary of the behavior of orbits of the logistic iteration rule for all relevant parameter values. It is also a lead-in to the next book in this series, *The Mandelbrot and Julia Sets,* because the orbit diagram is really the "spine" of the Mandelbrot set.

Mathematical Prerequisites

Students need to be familiar with the main points of Lessons 5, 6, and 8 in order to fully appreciate the structure of the orbit diagram.

Mathematical Connections

The major connection of this material to the secondary mathematics curriculum is the concept of **graphing dependent variables versus independent variables** and that such graphs do not have to use the typical y versus x Cartesian coordinate plane. The other important connection is an increasing awareness of **the process of investigation and experimentation in mathematics** to look for patterns, make conjectures, and categorize behavior. Students often feel that mathematics is simply a matter of memorizing rules and procedures established by mathematicians a long time ago and that there is nothing "new" left to discover. This lesson, the one to follow, and the fourth book in this series should make them feel otherwise.

TECHNOLOGY

Software to draw the orbit diagram is essential. The applet "Orbit Diagram" at our Web site **http://math.bu.edu/DYSYS/applets** draws all the orbit diagrams necessary to use this lesson. There is other freeware (such as Fractint) that also draws these images.

SUGGESTED LESSON PLAN

CLASS TIME

This material is ideally suited for use in a computer lab session. Or it can be assigned as an end-of-term project. Or it can be the subject of a one-hour summary class that recapitulates much of what has been covered before.

PREPARATION

Note that the orbit diagram is a very different image from those previously encountered. In this picture, the parameter c is plotted horizontally while the fate of the orbit—that is, the x-values—are plotted vertically. On the vertical line over each chosen c-value, we see displayed the fate of the critical orbit for that c-value.

Students should have performed (major portions of) Investigations 4–6 in Lesson 4 before, or at least concurrently with, this lesson.

LESSON DEVELOPMENT

It is important to stress that the orbit diagram is a picture of the fate of orbits, not the entire orbit. Any session dealing with this picture should begin by plotting a few points on the diagram as in the Explanation. Have students compute a few fates and draw them by hand on a displayed picture of the (blank) orbit diagram. Again, it may be necessary to discuss transient behavior and the fact that the first 25 points on the orbit are not plotted so as not to cloud the fate of the orbit with the transient behavior that occurs prior to the orbit settling in.

If a computer lab is available, turn students loose to experiment with the orbit diagram. It is a valuable experience for students just to zoom in and see the intricacy of this picture. Also, the self-similarity should remind them of fractal images from *Fractals,* the second book in this series.

A lab period can be structured by assigning Investigations 2 and/or 4.

Lesson Notes

Remember that a period n window is one that begins with an attracting period n cycle, followed by a $2n$-cycle, followed by a $4n$-cycle, and so forth. This means that the period 1 window contains all of the period 2, 4, 8, and subsequent cycles at the left end of the orbit diagram.

Note the power of the critical orbit: If the logistic iteration rule has an attracting fixed point or cycle, the orbit of the critical point tends to it. This is a good lead-in to the power of calculus, for the critical point is the place where the derivative vanishes. For students who plan to take calculus, this is a good concept to note at this stage.

Students sometimes ask: What if we use the orbit of some other seed to draw the orbit diagram? The answer is: We have no guarantee that the orbit of another seed will be attracted to an attracting cycle. This orbit may in fact land on a repelling cycle that may coexist with the attracting cycle. So, in the midst of a period n window, you may suddenly see "dirt" when the orbit happens to land on another, nonattracting orbit.

The Further Exploration problems in this lesson are really research projects. Each will challenge the best of students. Use these only for extra challenge or long-term group assignments. Do not expect immediate answers; do expect a lot of interesting and challenging questions. Answers such as "I think that so-and-so may be happening" are perfectly acceptable in this setting. The idea behind the problems is to get the students to think. Even if they do not come up with the right or even approximately right answer, if they have thought long and hard about the problem, that's fine!

In previous lessons, we discussed the logistic iteration rule, $x \rightarrow kx(1 - x)$. In this one family of iterations, we have seen a variety of phenomena, including cycles of many periods, chaos, and sensitivity to initial conditions. To understand these concepts, we used a number of different tools, including computer and calculator explorations, graphical iteration, time-series graphs, histograms, and more. In this lesson, we will summarize all that we have learned using a remarkable picture known as the **orbit diagram** (sometimes called the **bifurcation diagram**).

THE CRITICAL ORBIT

Recall that the orbit of the seed $x_0 = 0.5$ is called the critical orbit of the logistic iteration rule. In previous lessons, we saw that the critical orbit allowed us to see a pattern in the histograms of certain logistic iteration rules. This orbit has another special property: When a logistic iteration rule has an attracting cycle, then the critical orbit must tend to it. This means that a logistic iteration rule can have at most one attracting cycle, since an orbit cannot tend to two different cycles.

As we have seen, the logistic iteration rule $x \rightarrow 4x(1 - x)$ has cycles of many different periods. So at most, one of these cycles can be attracting. All the rest must be repelling. In fact, all the cycles for this iteration must be repelling because the critical orbit is eventually fixed at 0:

$$0.5 \rightarrow 1 \rightarrow 0 \rightarrow 0 \rightarrow 0 \ldots$$

and we know from graphical iteration that 0 is a repelling fixed point.

This means that the critical orbit provides an important way to discover the fates of orbits. If the iteration rule has an attracting cycle, the critical orbit finds it. If there are no attracting cycles, the histogram of typical orbits reflects the critical orbit.

We cannot prove that the critical orbit always "finds" attracting cycles. This demands many sophisticated mathematical ideas from calculus and a branch of mathematics known as *complex analysis.* But we can begin to appreciate the power of this fact at least experimentally.

Summary of logistic orbits

Let's now summarize all that we have learned about the logistic iteration rule $x \rightarrow kx(1 - x)$ for various k-values. We can do this effectively by finding the fate of the critical orbit. The easiest way to do this is to make a table of k-values and the corresponding fate of the critical orbit. Using what we have discovered in earlier lessons, we can create the following table:

k	**Fate of critical orbit**
0.5	tends to fixed point at 0
1.5	tends to fixed point at $\frac{1}{3}$
2.8	tends to fixed point at 0.643 . . .
3.2	tends to 2-cycle 0.513, 0.799
3.5	tends to 4-cycle 0.501, 0.875, 0.383, 0.827
3.6	appears chaotic
3.7	appears chaotic
3.83	tends to 3-cycle 0.505, 0.957, 0.156
3.9	appears chaotic
4	eventually fixed

This is a perfectly valid way to summarize the fate of orbits. However, we would need many more entries in this table to be able to capture the whole picture. A much more vivid way to do this is to paint this picture.

The orbit diagram

Instead of listing the fate of the critical orbit in a table, we will record the same information by plotting the k-values between 0 and 4 on the horizontal axis. (You will see why we choose a maximum value of 4 in the Investigations.) The vertical axis will consist of the x-values along the orbit. Above each given k-value, we will record the fate of the critical orbit. This means that we will plot only the final behavior of the orbit for each k; we will not plot any of the transient behavior. The resulting picture is known as an **orbit diagram.**

For example, in the preceding table we see that when $k = 0.5$, the critical orbit tends to a fixed point at 0. So we would plot the point (0.5, 0) indicating that when $k = 0.5$, the critical orbit tends to a single fixed point at 0.

When $k = 3.2$, the critical orbit tends to a 2-cycle

$$0.513 \rightarrow 0.799 \rightarrow 0.513 \rightarrow 0.799 \rightarrow \cdots$$

and so we would plot two points over $k = 3.2$: (3.2, 0.513) and (3.2, 0.799). When $k = 3.5$, the critical orbit tends to a 4-cycle, so we would plot those four points over $k = 3.5$. Here is the orbit diagram so far:

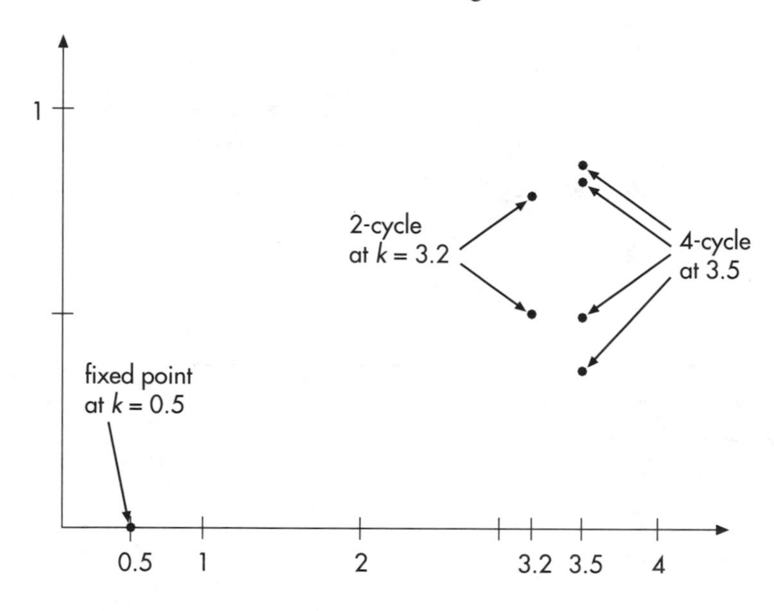

When $k = 3.9$, the orbit of 0.5 never settles down. It appears to wander aimlessly over an interval. We saw this best when we plotted a histogram of this orbit:

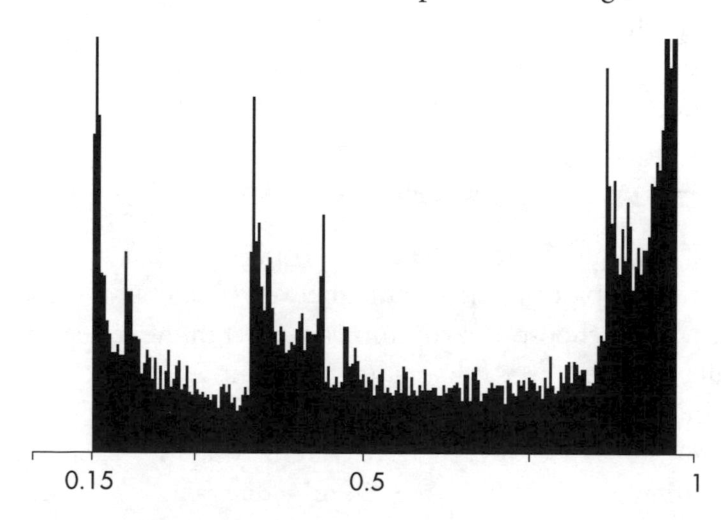

From this picture we see that the critical orbit seems to fill up the entire interval between approximately 0.15 and 0.95. So we would plot all of these points in the orbit diagram as an interval over $k = 3.9$. The orbit diagram now looks like this:

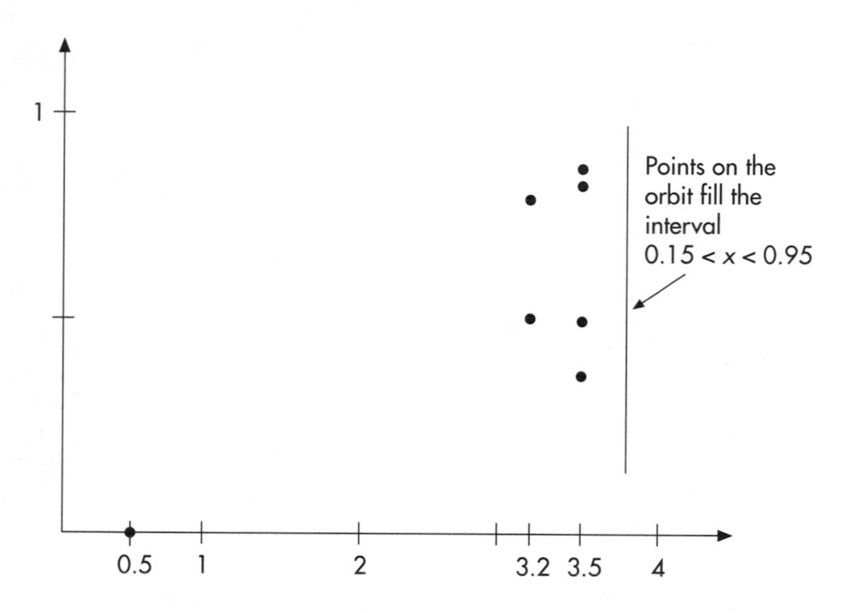

Now let's turn to the computer and let it do the work for us. In the following picture, we have selected 400 k-values between 0 and 4 (spaced at 0.01) and then computed 125 points on the orbit of $x_0 = 0.5$ for each k. We did not plot the first 25 points on this orbit; we plotted only the last 100 points so that we remove the transient behavior.

Here is how the computer paints the orbit diagram:

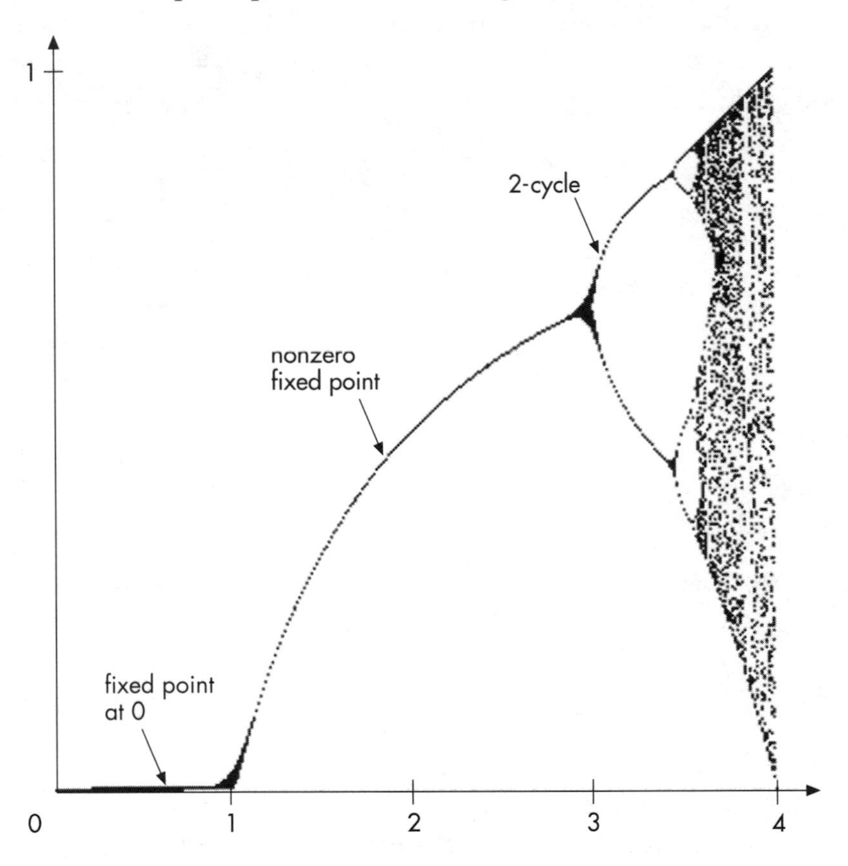

Note that we can read off immediately from this picture some of the observations we have made in earlier lessons. When k is between 0 and 1, the orbit seems to tend to a fixed point at 0. When k is between 1 and approximately 3, it appears that the critical orbit tends to a nonzero fixed point. For larger values of k, it appears that the critical orbit tends to a 2-cycle. Then the picture seems to get cloudy. So let's look more closely at this portion of the orbit diagram by magnifying the interval $3 \leq k \leq 4$:

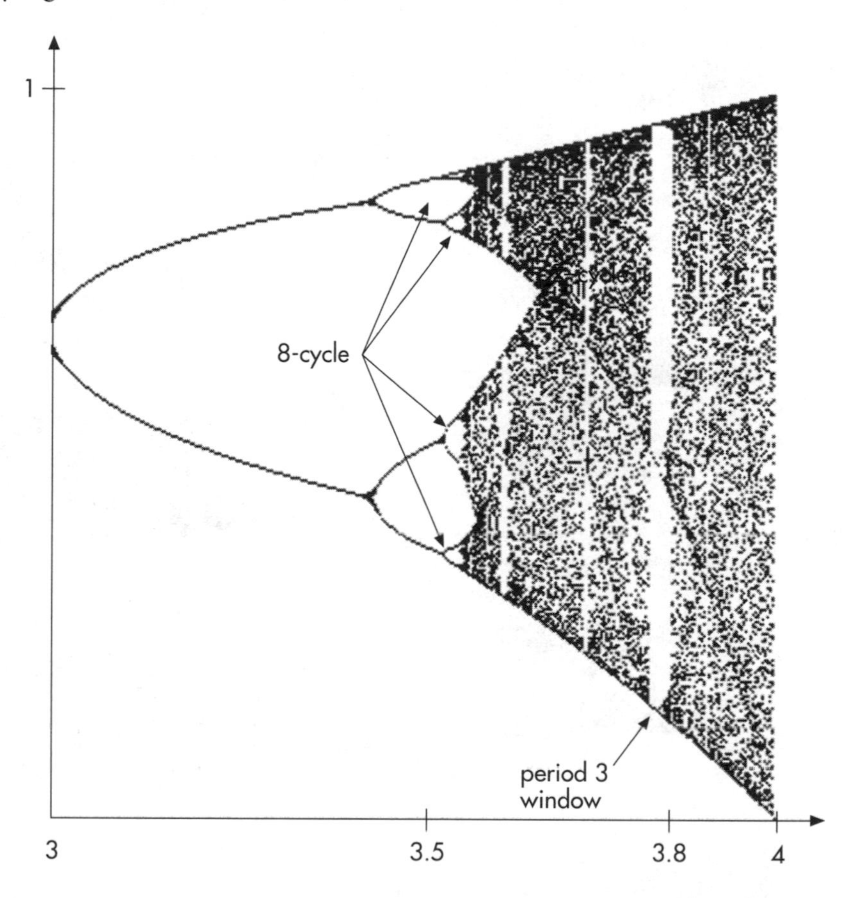

In this magnification, we see even more detail. The 2-cycle gives way to a 4-cycle as k increases, and then the 4-cycle seems to give birth to an 8-cycle. For larger k-values, it appears that there is no pattern to the orbit as it fills an interval. This collection of k-values is the chaotic regime. However, there are several exceptions. For k-values slightly larger than 3.8, it appears that a "window" opens where there is no chaos. Instead, we seem to see only a cycle of period 3.

THE PERIOD 3 WINDOW

Let's investigate this portion more closely. In this picture, we have plotted the portion of the orbit diagram over the interval $3.81 \leq k \leq 3.88$:

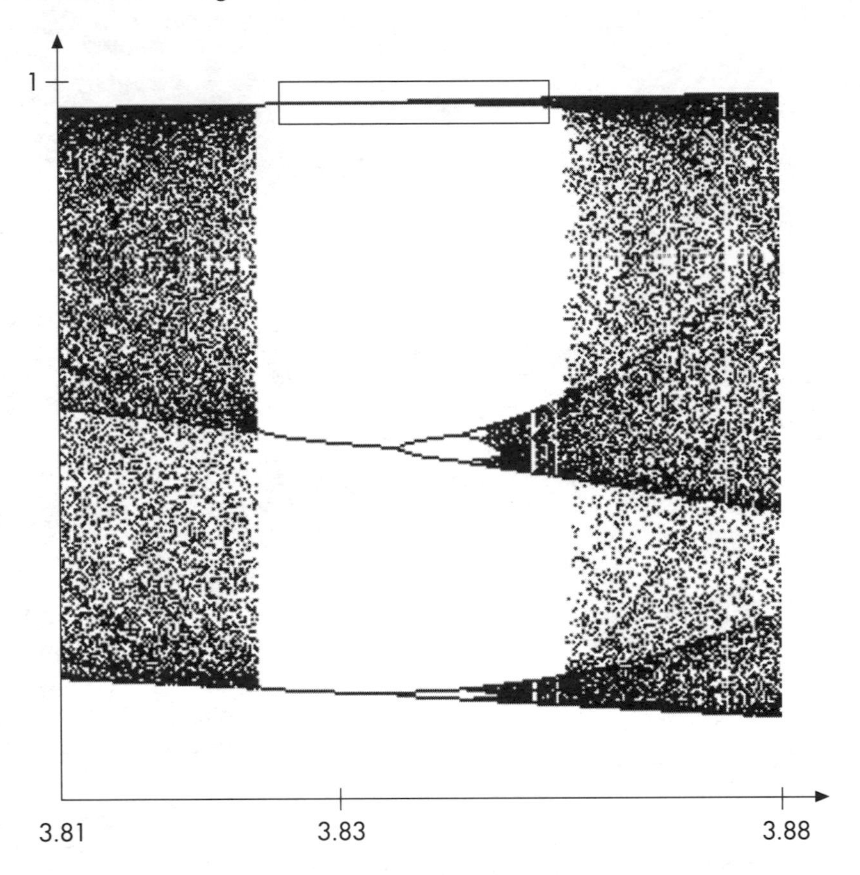

Now we see something remarkable. In the midst of the chaotic regime, suddenly the critical orbit ceases to be chaotic. Instead, it becomes attracted to a 3-cycle. Then that 3-cycle seems to double in period just like our original fixed point did. Let's see that by magnifying the small box we have outlined in the period 3 window:

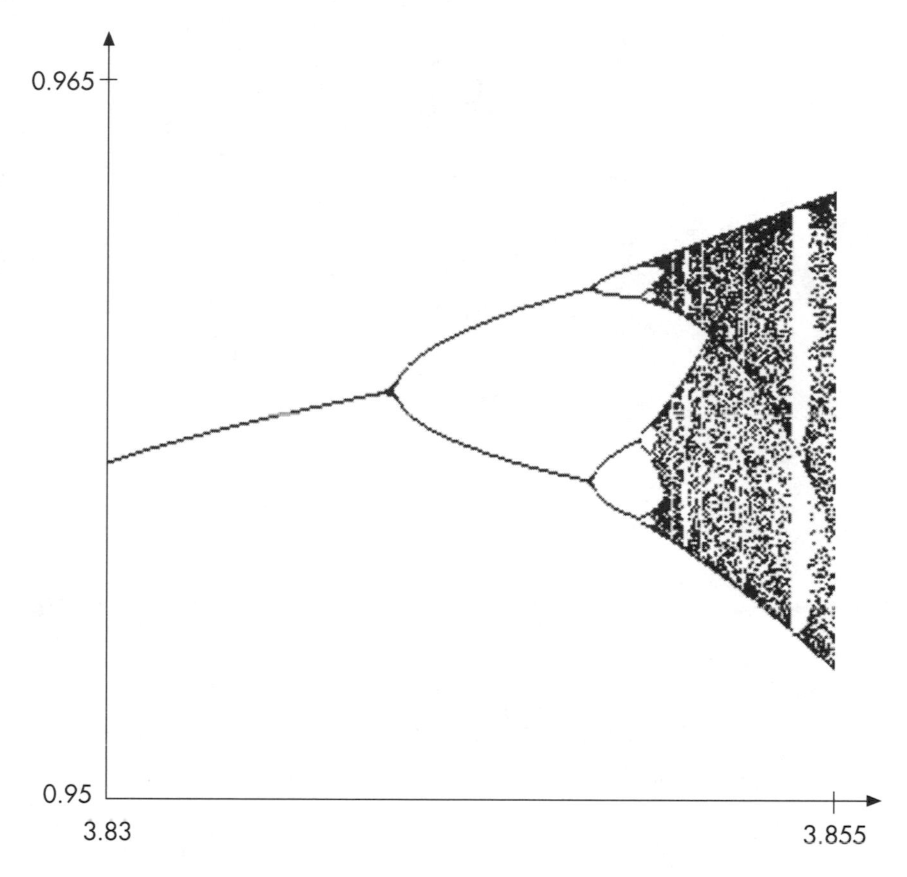

In the period 3 window, we find three smaller copies that look very similar to a portion of the original orbit diagram! This is reminiscent of the notion of self-similarity that you saw if you read *Fractals,* the second book in this series.

The period-doubling route to chaos

The orbit diagram shows us that many amazing things happen as k increases in the logistic iteration rule. First we see an attracting cycle. Then its period is doubled and we see a 2-cycle. Then the period doubles again and we see a 4-cycle. Then an 8-cycle. And on and on until the period becomes infinite and we enter the chaotic regime. This is called the **period-doubling route to chaos.**

A glance at the period 3 window shows that the same period-doubling occurs there. First a 3-cycle, then a 6-cycle. Then 12. Then 24. And on and on to the chaotic regime.

But there is more to this story than just period-doubling. Look closely at the original orbit diagram and several successive magnifications:

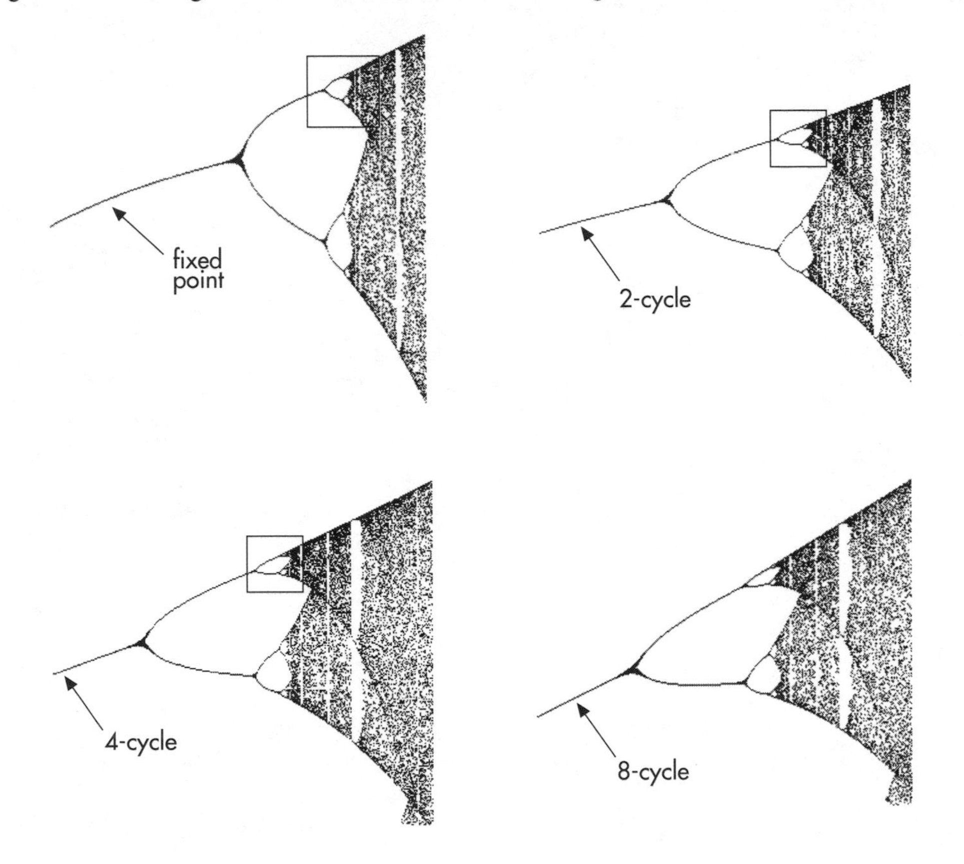

Note how portions of these magnifications resemble the original picture. We see a succession of period doublings, followed by a chaotic regime, then a period 3 window (actually, a piece of a $3 \cdot 2^n$ window). Everywhere we look inside these windows we see smaller and smaller copies of the same picture. You might be tempted to call this picture a fractal since there is clearly some self-similarity. But recall that there is another element to the definition of a fractal, namely, fractal dimension. These pieces of the orbit diagram have certain portions that are one-dimensional, and other portions that seem to have higher dimension. So strictly speaking, the orbit diagram is not a fractal. Nevertheless, it certainly does exhibit some self-similarity.

APPLICATIONS OF PERIOD-DOUBLING

Many real-world systems that progress from simple to complicated behavior do so via the period-doubling route to chaos. For example, certain hydrodynamic systems, when heated, exhibit successive changes that involve fluids moving around in circular patterns whose period doubles and doubles as the temperature rises. Similarly, scientists have observed period-doubling in the formation of weather patterns, in oscillating chemical reactions, and in electrical circuits.

1 ▹ LARGER k-VALUES

When we plotted the orbit diagram for the logistic iteration rule, we used only k-values that were less than 4. Why? What can you say about the critical orbit when $k > 4$? Use graphical iteration to support your idea.

2 ▹ FINDING WINDOWS

By experimentation, find windows in the orbit diagram of periods 4 through 10. Remember that a window must feature initial period n, then $2n$, then $4n$, and so forth. You may wish to use the applets located in **http://math.bu.edu/DYSYS/applets** to do this. List the approximate k-interval containing your window. *Hint:* There are many possible answers.

3 ▹ WORLD'S LONGEST ORBIT DIAGRAM

By using a computer to draw portions of the orbit diagram over small k-intervals, piece together a long image of the orbit diagram. Indicate on this diagram the locations of various period n windows.

4 ▹ Windows between windows

Note that there are two windows between the period 1 window (that's the one that contains the fixed points and all of its period-doublings) and the period 3 window. Actually, there are infinitely many such windows, but there are two that are the widest in terms of k-values.

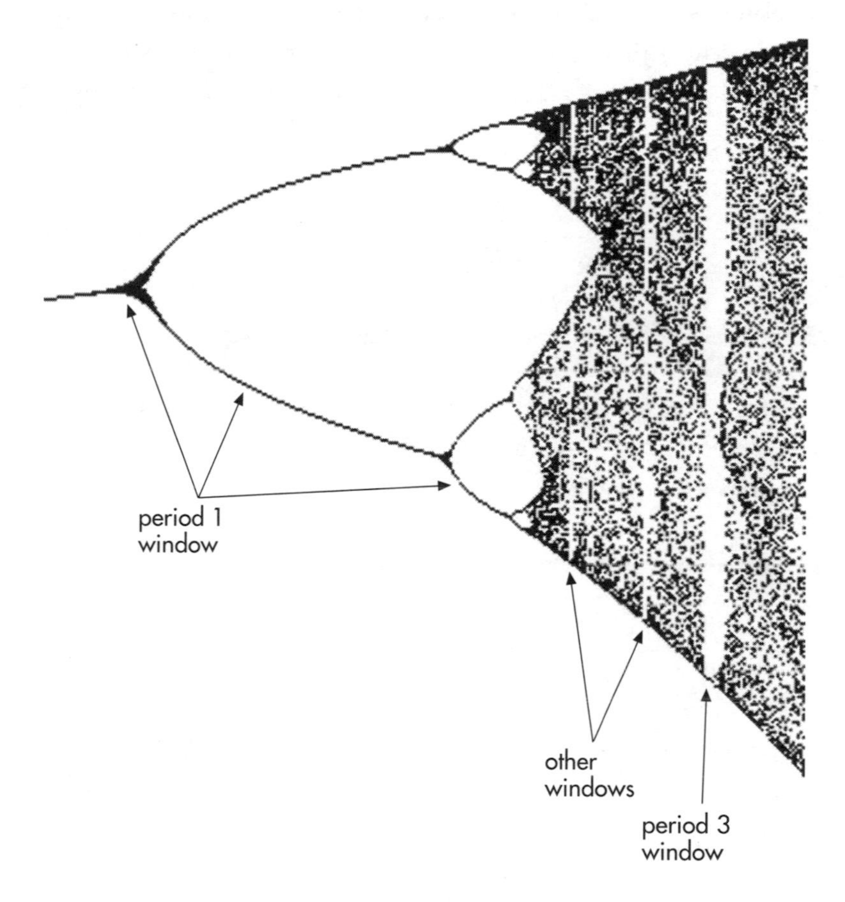

a. What are the periods of these windows (labeled "other windows" in the diagram)?

b. Now magnify the region between the period 1 window and the leftmost window you just found. What are the periods of the widest two windows in this region?

____

____

c. Now do this one more time. Magnify the region between the new leftmost window and the period 1 window. What are the periods of the two widest windows in this region? Do you see a pattern beginning to form?

____

____

5 ▹ Smears in the Orbit Diagram

In certain areas of the orbit diagram, there are "smears" or clouds of points, for example, near $k = 1$ and $k = 3$. Using any tool you have available, give an explanation of what causes these smears.

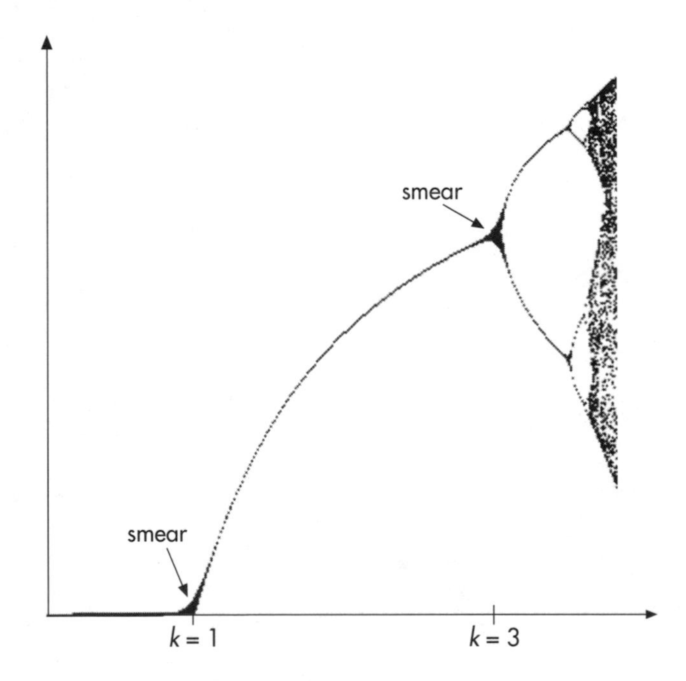

1. Use a computer, spreadsheet, or the applets listed earlier to compute the orbit diagram for the iteration rule

$$x \to k\left(x - \frac{x^3}{3}\right)$$

Let k vary in the interval $0 \le k \le 2.598\ldots$ and view the orbits in the interval $0 \le x \le \sqrt{3}$. The critical orbit here is the orbit of 1 since this function achieves its maximum value at $x = 1$. Compare this orbit diagram to the orbit diagram for the logistic iteration rule.

2. Here are the iteration rule and its graph. This rule is called the "tent map" because of the shape of the graph:

$$x \to \begin{cases} cx & \text{if } 0 \le x < \frac{1}{2} \\ c(1 - x) & \text{if } \frac{1}{2} \le x \le 1 \end{cases}$$

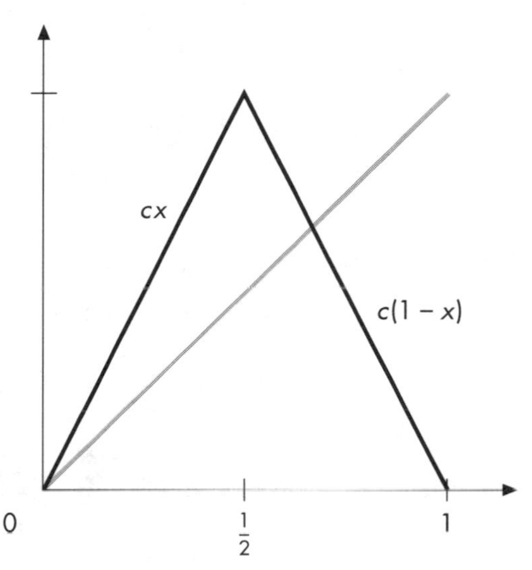

At right is the orbit diagram for this iteration rule. In this image, the critical orbit is the orbit of ½. The values of c should run from 0 to 2, and the orbits are plotted in the interval $0 \le x \le 1$. Can you figure out why the "hole" appears in this orbit diagram? And what change occurs at $c = 1$?

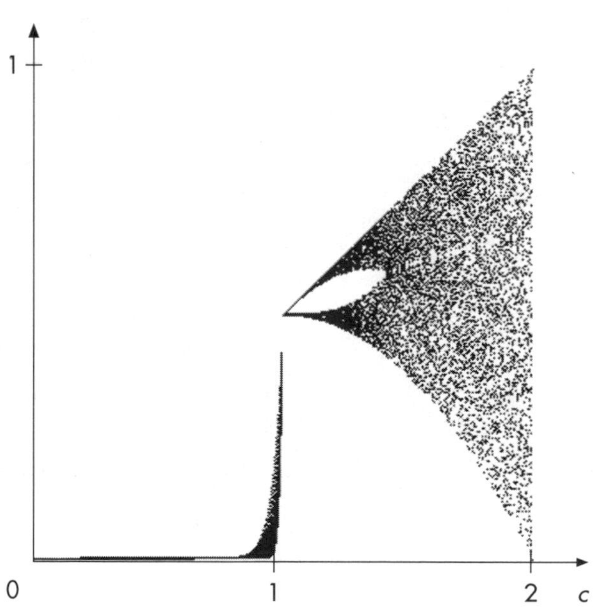

3. Consider the iteration rule $x \rightarrow A\cos(x)$. Here the parameter A runs from 1 to 6.2. The critical orbit is the orbit of $-\pi$ since the graph of $y = A\cos(x)$ has a minimum value at this x-value. (You can also use the orbit of 0 as a critical orbit since $y = A\cos(x)$ has a maximum value at 0.) In the orbit diagram, we have plotted the parameter A on the horizontal axis and the x-values along the orbit on the vertical axis.

   The left portion of this orbit diagram looks familiar, but something happens when $A = 2.975\ldots$. Using graphical iteration, try to explain what happens at this A-value. There is another change near $A = 4.19\ldots$. Again use graphical iteration to explain this sudden change in the orbit diagram.

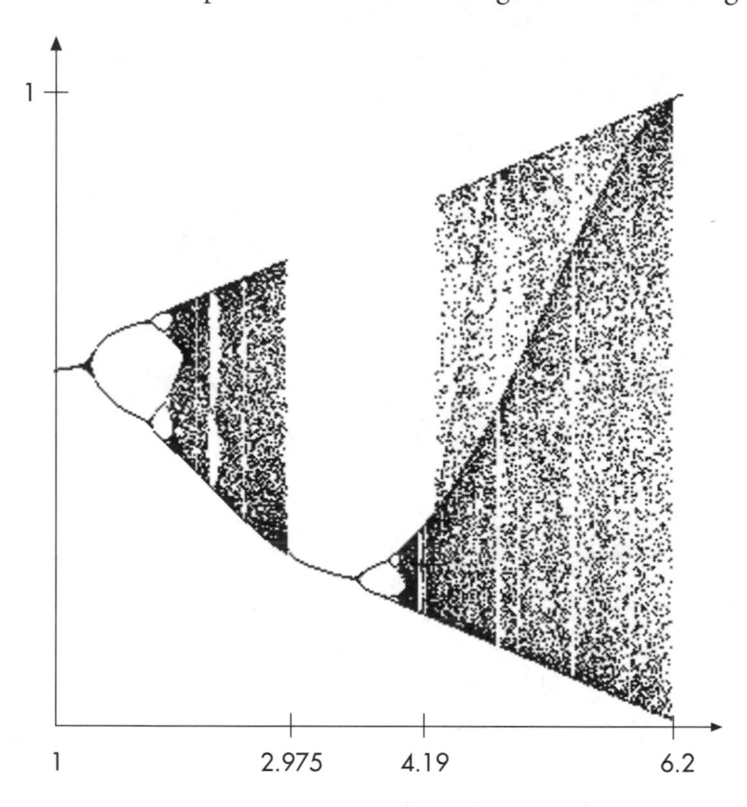

4. Use a computer or the applets mentioned earlier to plot the orbit diagram of the iteration rule $x \rightarrow A\sin(x)$ for A-values in the interval $0 \leq A \leq \pi$ and x-values satisfying $0 \leq x \leq \pi$. Use the orbit of $\pi/2$ as the critical orbit. Compare this orbit diagram to the logistic orbit diagram.

Teacher Notes

A Quadratic Expedition

Overview

This lesson is a collection of investigations dealing with the quadratic iteration rule. It can be assigned as a final project or used as a lead-in to the fourth book in this series, *The Mandelbrot and Julia Sets,* which deals with the complex quadratic iteration rule.

Mathematical Prerequisites

Students need to be familiar with much of the preceding material in this book in order to understand the Investigations.

Mathematical Connections

The Mandelbrot set and Julia sets for quadratic iteration rules make heavy use of the material in this lesson. This material provides a solid mathematical foundation for investigating these topics.

As in the previous lesson, one of the most important connections for students is to make them aware that mathematics is not simply "black and white" as they may have been led to believe. It does not always consist of right or wrong answers to a set of rules and procedures that were all discovered a long time ago. It is a field full of investigation, experimentation, and research that has undergone a rebirth with the advances in technology. Most students have never given much thought to what mathematicians "do."

Technology

A graphing calculator is sufficient for all Investigations in this lesson.

SUGGESTED LESSON PLAN

CLASS TIME

Assign this lesson as a term project or final project for students who have mastered the material in this book.

LESSON NOTES

In Investigation 1, it is difficult to determine experimentally the exact c-values for which the orbit of 0 tends to infinity or, as we say in this lesson, escapes. However, graphical iteration, together with the results of Investigation 2, comes to the rescue, for the orbit of 0 escapes when there are no fixed points and when the value of c lies below the eventual fixed point on the negative axis. *Note:* The quadratic formula and the value of the discriminant allow students to algebraically determine the number of fixed points in Investigation 2.

Again, in Investigation 3 it is difficult to determine the exact c-values for which there is an attracting fixed point. Finding an approximate range in which there is an attracting fixed point is sufficient. The same comment holds true for Investigation 6 where it is difficult to determine the exact value for which the 2-cycle emerges without doing the algebra.

In previous lessons, we have concentrated on the logistic iteration rule, $x \rightarrow kx(1 - x)$. In this iteration rule, we have found a great variety of different phenomena and we have used a variety of tools to understand these topics. In this lesson, we will continue to use these ideas to explore another iteration, the quadratic iteration rule, $x \rightarrow x^2 + c$. Here, c is a parameter, just as k was in the logistic iteration rule. As we will see in the next book in this series, *The Mandelbrot and Julia Sets,* this is the iteration rule that produces the famous Mandelbrot set. So this lesson can be regarded as a warm-up for this very interesting and timely topic. It is also a warm-up in another sense. Here we will let you do all the work; this entire lesson is one long exploration into the behavior of the quadratic iteration rule.

WHAT MATHEMATICIANS DO

Whenever a mathematician encounters a problem, the first thing he or she does is try a few cases, experiment a bit. In years past, this often involved long computations done by hand, or elaborate graphs, again painstakingly drawn by hand. Now we have many different and much more powerful tools, including the computer, which handles the long and tedious calculations with ease, and the graphing calculator, which displays accurate graphs in seconds. This technology has changed the playing field dramatically for mathematicians. It has opened up whole new areas for experimentation, and it has shown us images that we could never before have imagined.

Once the mathematician has gained some insight into the problem experimentally, he or she then brings out the mathematical toolbox that includes algebra, geometry, trigonometry, and the like to try to understand or solve the problem as accurately and completely as possible.

In this lesson, you are the mathematician. We will lead you through a series of open-ended Investigations aimed at understanding the quadratic iteration rule. There is not always one correct answer here, for these Investigations may lead you in many different directions. If you find something that intrigues you, by all means make a detour and investigate this topic further. If you find something that stumps you, don't worry. Mathematicians have tried for years to understand the quadratic iteration rule completely. Although much progress has been made, the full story is still not known. You may even discover something about this family of rules that has not been known or seen before!

THE ORBIT OF 0

In this lesson, we will concentrate on the orbit of 0. This orbit is the critical orbit for the quadratic iteration rule since 0 is the place where the expression $x^2 + c$ achieves its minimum, or critical, value.

1 ▹ NON-ESCAPE *c*-VALUES

Let's first determine the *c*-values for which the orbit of 0 does not go to infinity. In order to do this, you should use a variety of techniques, including graphical, numerical (calculator or computer), and algebraic. Here are some graphs of special cases to get you started:

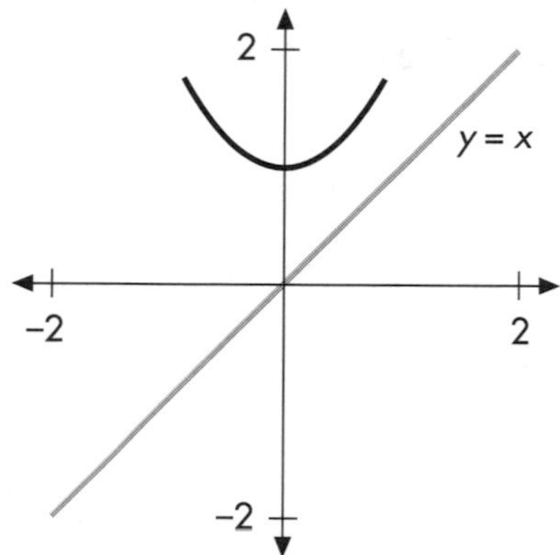

$y = x^2 + 1$

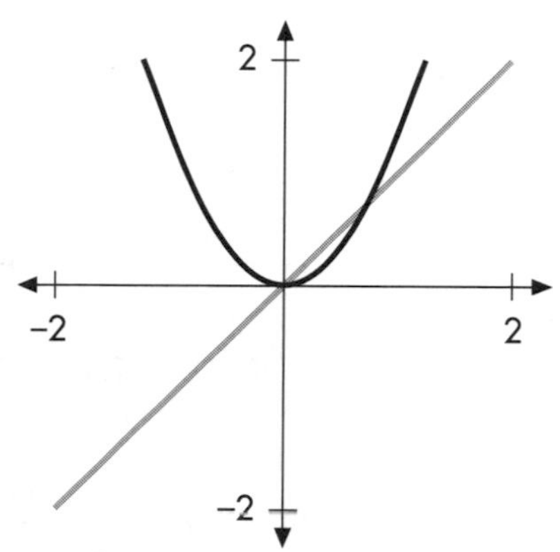

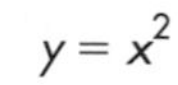

$y = x^2$

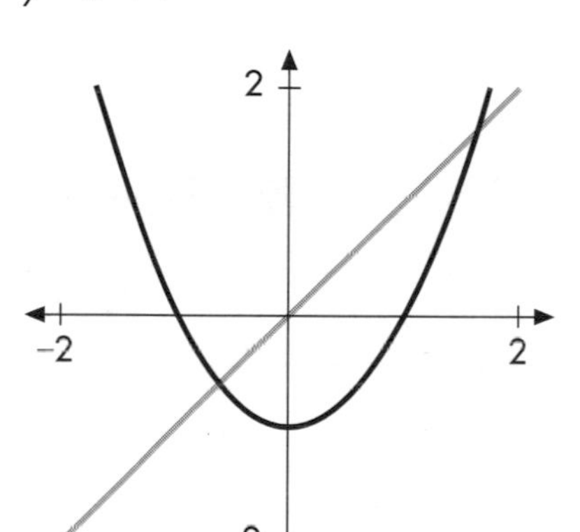

$y = x^2 - 1$

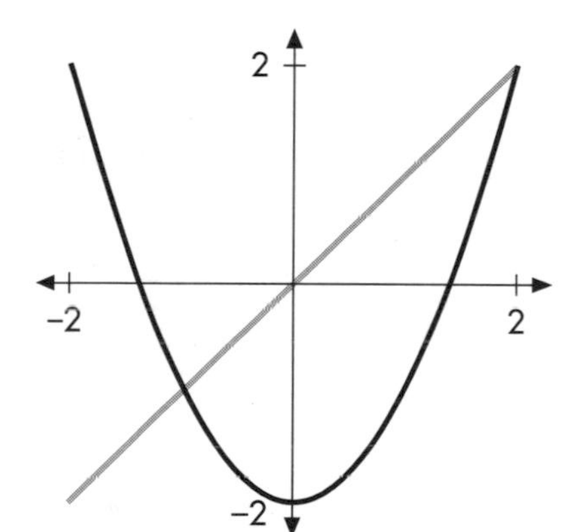

$y = x^2 - 2$

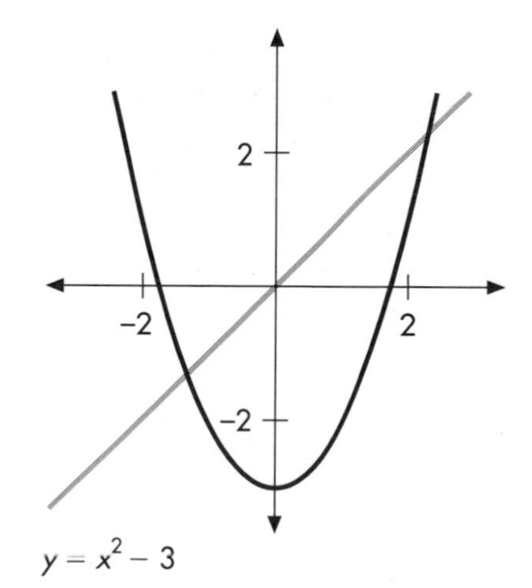

$y = x^2 - 3$

a. A glance at the graph of $y = x^2 + c$ for the c-values given above and some graphical iteration should help you start to look for the c-values for which the orbit of 0 does not go to infinity. Use the computer or calculator to find the c-values more explicitly. *Note:* Don't just use integer c-values; use intermediate values such as $c = -1.567$ or $c = -2.01$. Then, using algebra, determine the exact c-values for which the orbit of 0 does not go to infinity.

The c-values for which the orbit of 0 does not go to infinity are:

b. Briefly describe how you found these c-values. Which experiments did you perform? What algebra did you do?

2 ▸ FIXED POINTS

Find the values of c for which the iteration rule $x \rightarrow x^2 + c$ has fixed points. For which c-values are there no fixed points?

Only one fixed point?

More than one fixed point? How many fixed points?

How do your results in this Investigation relate to the results of the previous Investigation?

3 ▸ ATTRACTING FIXED POINTS

For which *c*-values does the iteration rule $x \to x^2 + c$ appear to have an attracting fixed point? Use both a calculator or a computer and graphical iteration to explain these results.

4 ▸ OTHER ORBITS

So far we have looked at only the orbit of 0 for this iteration rule. Now let's look at all other orbits, at least in some special cases. Use graphical iteration to understand the fate of all orbits for each of the following iteration rules. First sketch the different fates using different colors. Then describe your conclusions in words.

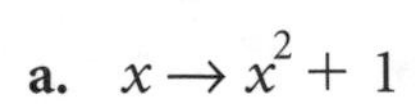

a. $x \rightarrow x^2 + 1$

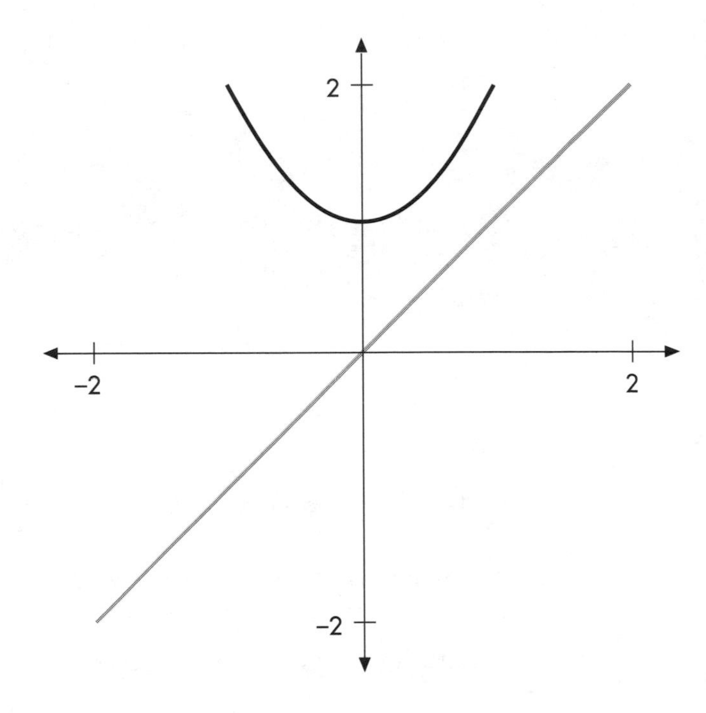

____

____

b. $x \rightarrow x^2$

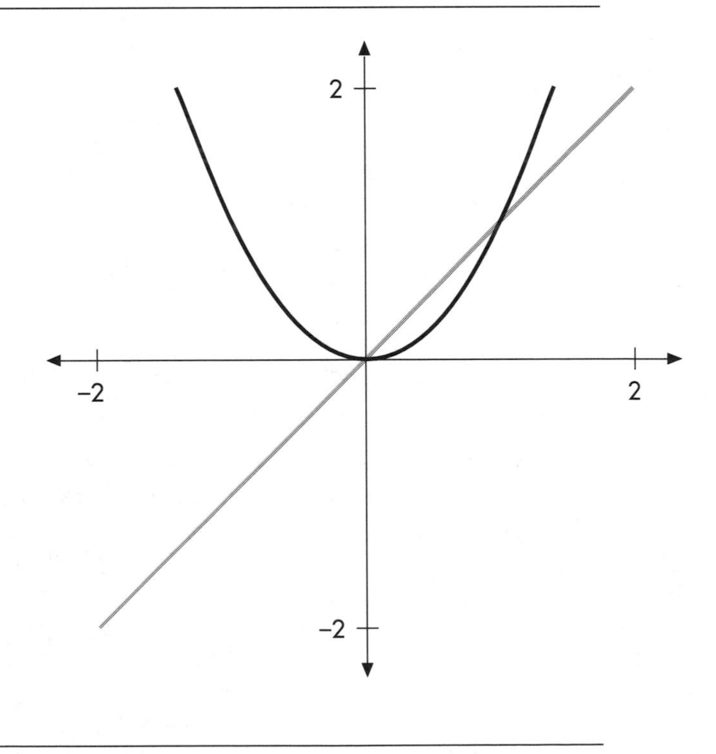

____

____

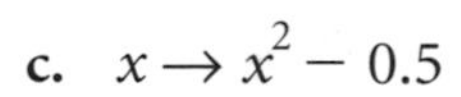

c. $x \to x^2 - 0.5$

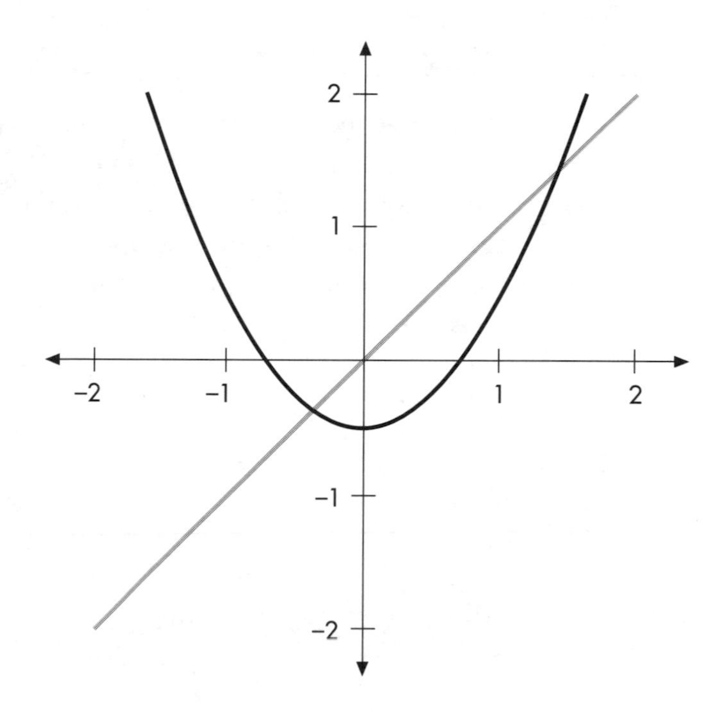

____

____

d. $x \to x^2 - 1$

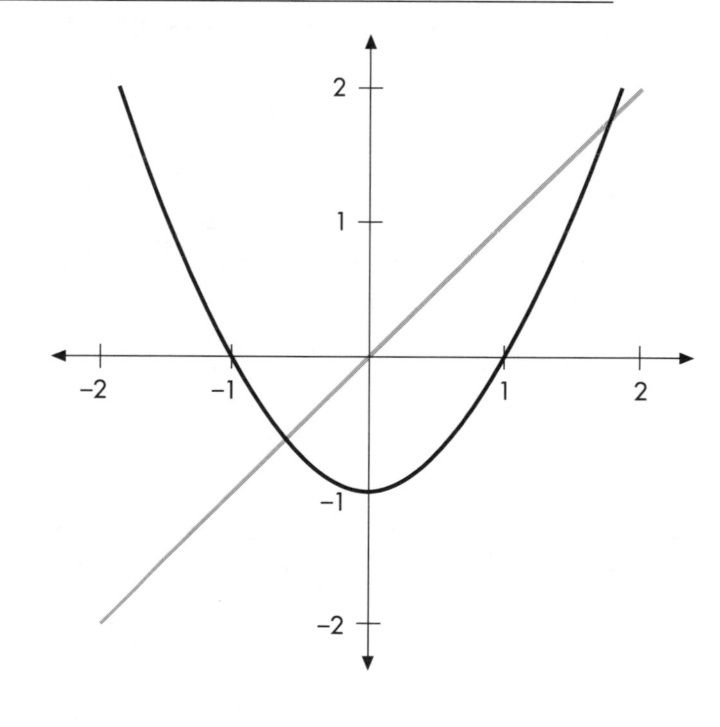

____

____

5 ▹ NON-ESCAPE ORBITS

Whenever the orbit of 0 goes to infinity, we say that the orbit escapes. Whenever the orbit of 0 does not escape, there is always an interval of seeds whose orbits do not go to infinity. First, using the results of Investigation 1, find this interval of non-escaping seeds exactly. Then explain how you found this interval using a picture of the graph of $x \rightarrow x^2 + c$ as well as your own words. What, in particular, can you say about the endpoints of this interval?

____

____

____

6 ▹ WHAT HAS CHANGED?

From the results of Investigation 4, something has changed in the fate of the orbit of 0 as we varied the parameter c from 0 to -1.

a. By choosing many other c-values between 0 and -1, determine as best as you can where this change takes place.

____

____

b. Sketch the graphs of $y = x^2 + c$ for several c-values just before, at, and just after this change occurs. Also include graphical iteration showing the fate of the orbit of 0 in each case.

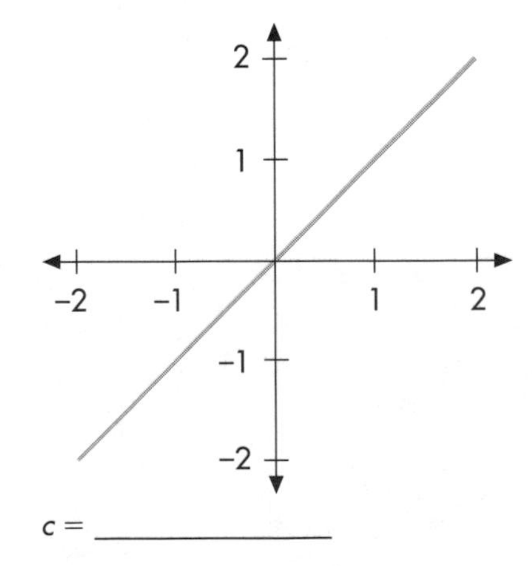

$c =$ ____________

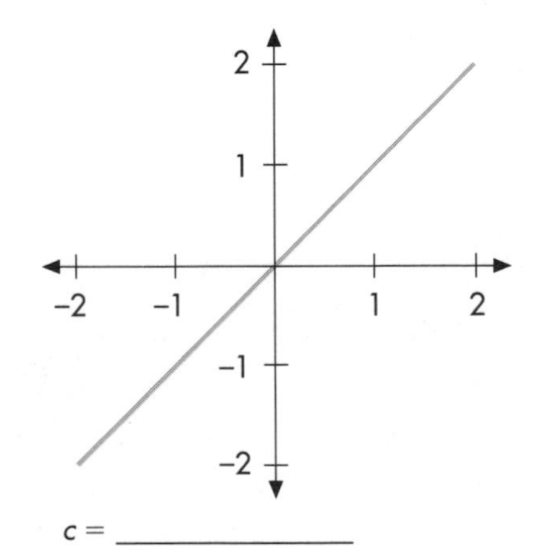

$c =$ ____________

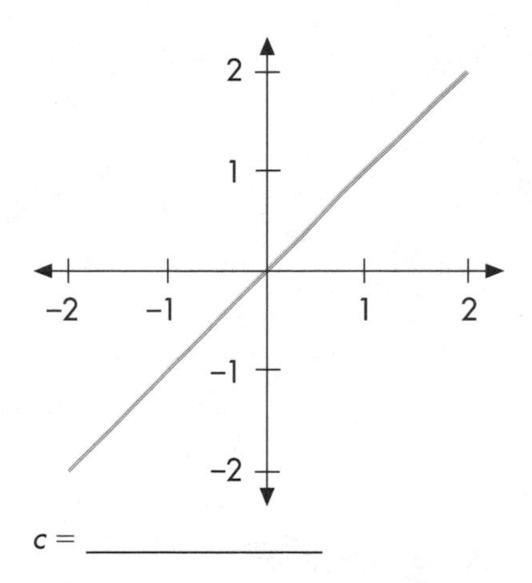

$c =$ ______________

c. Give the formula for the second iteration of $x \rightarrow x^2 + c$.

____

____

d. Then, using a graphing calculator, draw and record the graph of this expression for each of the three *c*-values you used in part b.

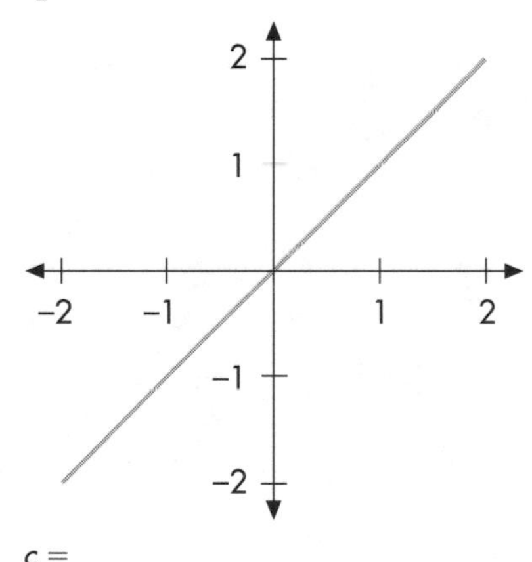

$c =$ ______________

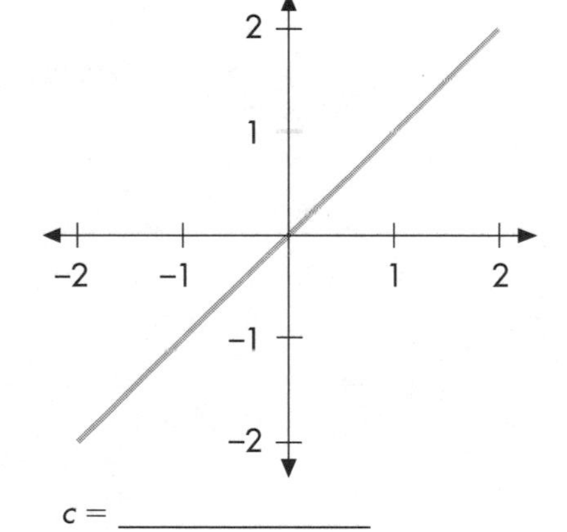

$c =$ ______________

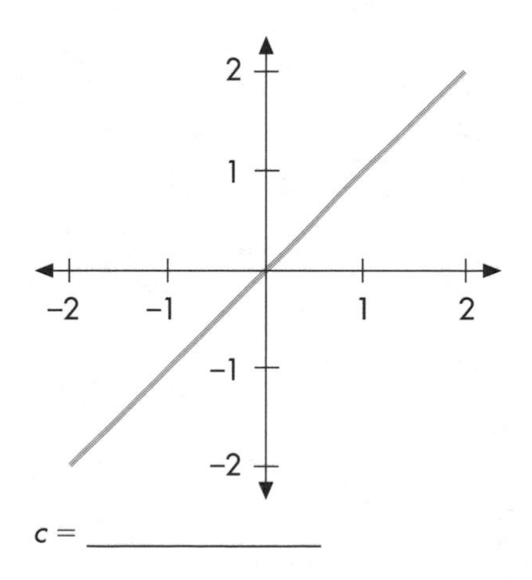

$c =$ ______________

e. In a brief essay, discuss the relevant features of these graphs as well as the change that has taken place as you lower the *c*-value.

7 ▸ OTHER *c*-VALUES

a. Using a calculator or a computer, investigate the fate of the orbit of 0 for a large number of *c*-values less than -1 (excluding those for which the orbit of 0 goes to infinity). If the orbit of 0 tends to a cycle, record both the period and the actual orbit of this cycle. If the orbit of 0 appears to be chaotic, write "chaotic" next to that *c*. Your investigations should include at least the following *c*-values:

$c = -1.3$

$c = -1.35$

$c = -1.4$

$c = -1.5$

$c = -1.77$

$c = -1.8$

$c = -2$

b. Using the data you have collected, display the fate of the orbit of 0 in an orbit diagram as described in Lesson 9. Remember that this is a plot of (just) the fate of the orbit of 0 versus the parameter c. So we record the c-value along the horizontal axis and the orbit (maybe a fixed point or a cycle) on the vertical line over the relevant c. *Note:* We have deliberately left off the numbers on both axes here. Using your results of the previous Investigations, you should be able to determine the relevant size scales of these axes.

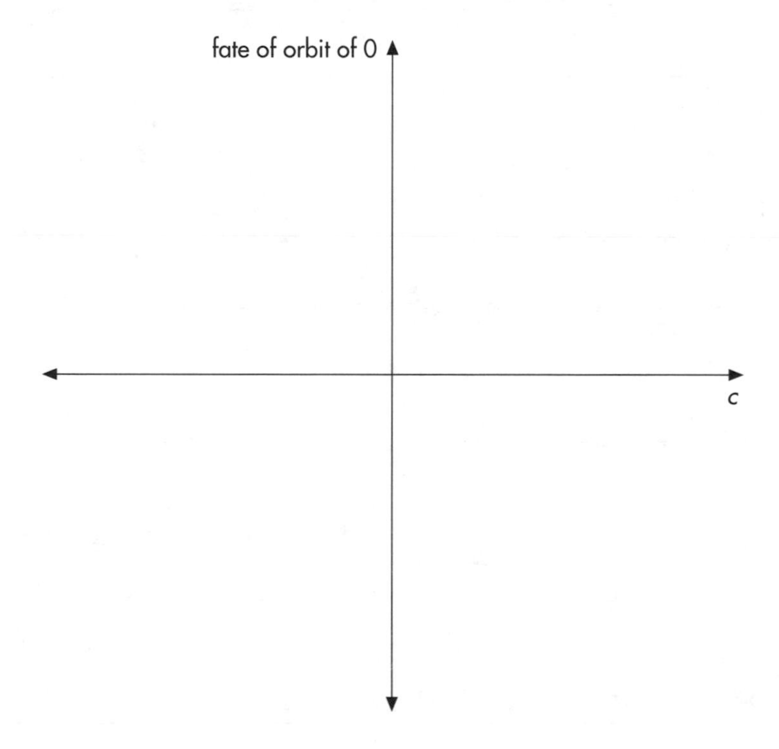

c. How does your picture relate to the orbit diagram for the logistic iteration rule?

___

___

8 ▹ A SPECIAL CASE

The case is $c = -2$. Let's spend some time discussing a particular case of the quadratic iteration rule, namely, $x \rightarrow x^2 - 2$.

a. What are the fixed points for this rule?

___

___

b. What is the fate of the orbit of 0?

____

____

c. Which orbits go to infinity and which do not in this case?

____

____

d. Does this iteration rule exhibit sensitivity to initial conditions? Discuss the fate of several orbits to explain your answer.

____

____

____

____

____

e. Display a histogram, time series, and graphical iteration for a "typical" orbit of $x \to x^2 - 2$. (Don't be coy and simply give the fixed point or something like that!)

9 ▸ Graphing Higher Iterations

a. Use a graphing calculator to draw the graphs of the first, second, third, and fourth iterations of $x \to x^2 - 2$ over the interval where orbits do not escape, and then record your results. How many points of intersection does each of these graphs have with the diagonal?

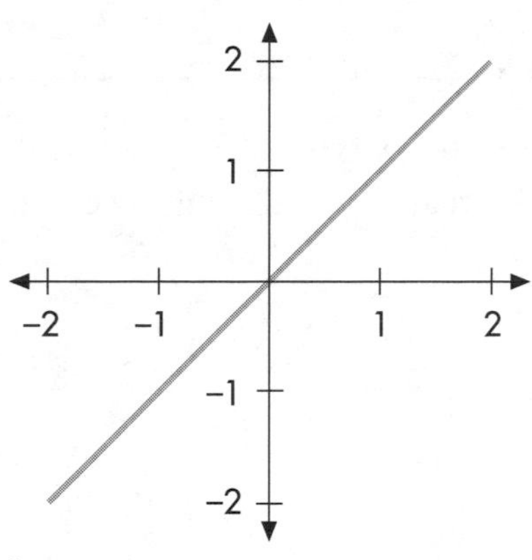

first iteration ______________

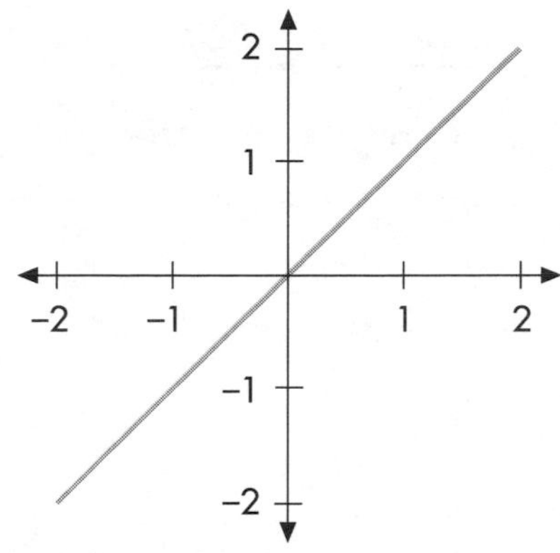

second iteration ______________

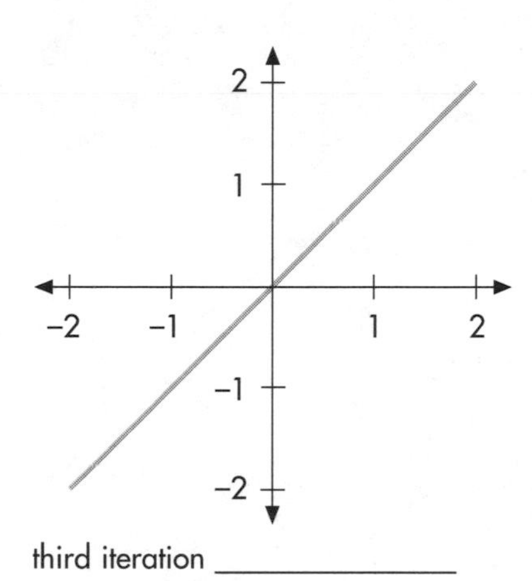

third iteration ______________

fourth iteration ______________

b. Can you draw any conclusions about subsequent iterations?

____

____

____

____

10 ▹ TARGET PRACTICE

a. Can you find other cycles for this iteration rule? Use graphical iteration and our target practice method to find a 3-cycle, 4-cycle, and 5-cycle using the graph below.

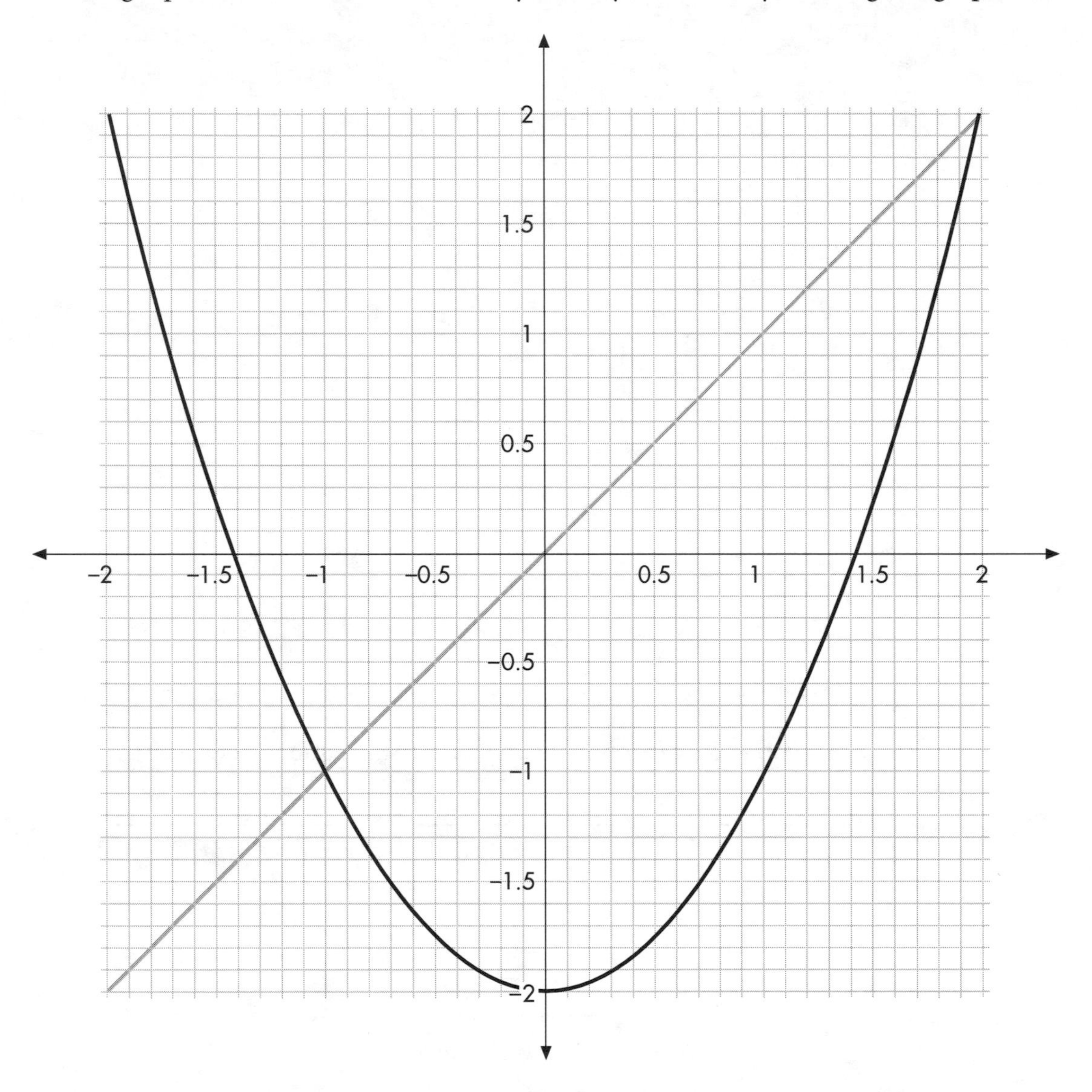

b. **Summary:** Using a combination of graphs, time series, histograms, orbit diagrams, and so on, write a three-page summary of all that you have found regarding the iteration rule $x \rightarrow x^2 + c$. Be sure that your report is in proper English with good sentence and paragraph structure. Include only relevant figures in your essay.

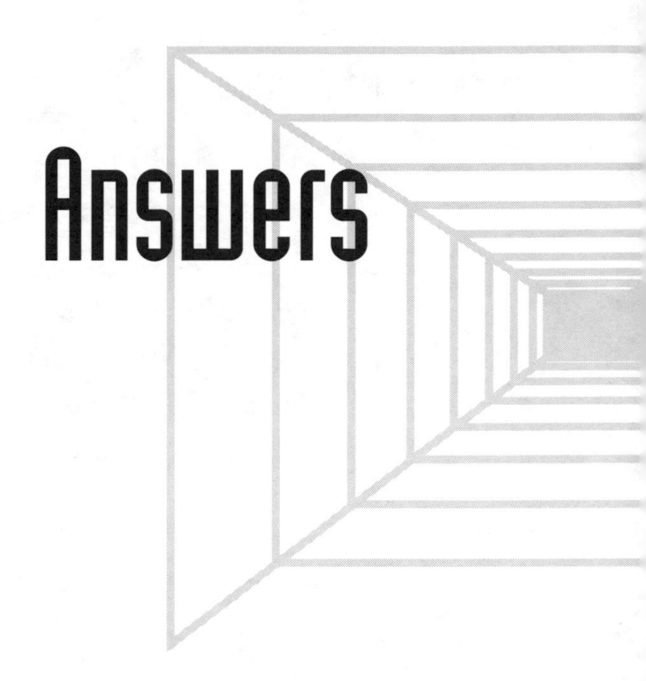

Answers

Lesson 1 ▸ Linear Iteration

Investigation 1: Computing orbits

a. For every seed the orbits tend to 4, the fixed point.

b. All orbits go to 4.

c. Possible answer: Seeds of 3.7, 2.2, and −2.2 go to 4.

d. Yes

e. The time series below was created using Excel.

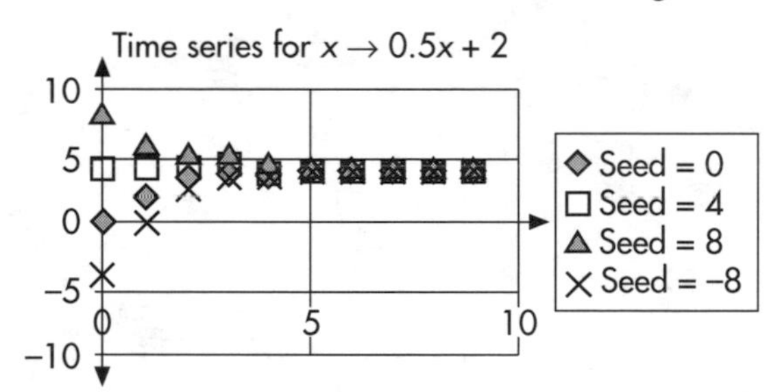

INVESTIGATION 2: ANOTHER ORBIT

a. The orbits of 3 and 5 tend to infinity. The orbits of 0, 1, −5, and −2 tend to negative infinity. The orbit of 2 is fixed.

b. If the seed is 2, the orbit is fixed. If the seed is less than 2, the orbits go off to negative infinity. If the seed is greater than 2, the orbits go off to infinity.

c. There are three different types of orbits.

d. The orbit of the seed 2 is fixed.

e. Here are the time-series plots for $x \rightarrow 2x - 2$:

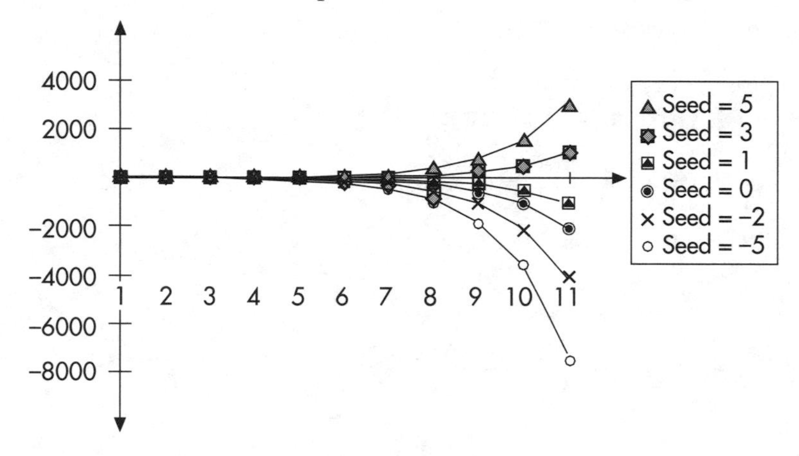

INVESTIGATION 3: ONE MORE ORBIT

a. The fixed point is $x = 1$.

b. In this case, the orbits are all repelled from the fixed point but they alternate in sign.

n	Orbits						
0	−1	10	−6	−10	−0.25	0.125	2
1	5	−17	15	23	3.5	2.75	−1
2	−7	37	−27	−43	−4	−2.5	5
3	17	−71	57	89	11	8	−7
4	−31	145	−111	−175	−19	−13	17
5	65	−287	225	353	41	29	−31
6	−127	577	−447	−703	−79	−55	65
7	257	−1151	897	1409	161	113	−127
8	−511	2305	−1791	−2815	−319	−223	257
9	1025	−4607	3585	5633	641	449	−511
10	−2047	9217	−7167	−11263	−1279	−895	1025
11	4097	−18431	14337	22529	2561	1793	−2047
12	−8191	36865	−28671	−45055	−5119	−3583	4097
13	16385	−73727	57345	90113	10241	7169	−8191
14	−32767	147457	−114687	−180223	−20479	−14335	16385
15	65537	−294911	229377	360449	40961	28673	−32767

c. All orbits except those of the fixed point get larger and larger in absolute value and alternate in sign.

INVESTIGATION 4: STILL ANOTHER ORBIT

a. The fixed point is 2.

b. The table of orbits given below shows that all the orbits tend to the fixed point 2.

n	Orbits						
0	0	–4	10	4	–10	100	–46.3
1	3	5	–2	1	8	–47	26.15
2	1.5	0.5	4	2.5	–1	26.5	–10.08
3	2.25	2.75	1	1.75	3.5	–10.25	8.0375
4	1.875	1.625	2.5	2.125	1.25	8.125	–1.01875
5	2.0625	2.1875	1.75	1.9375	2.375	–1.0625	–3.50938
6	1.9688	1.90625	2.125	2.0313	1.8125	3.53125	1.24531
7	2.0156	2.04688	1.9375	1.9844	2.0938	1.23438	3.37734
8	1.9922	1.97656	2.03125	2.0078	1.9531	2.38281	1.81133
9	2.0039	2.01172	1.98438	1.9961	2.0234	1.80859	2.09434
10	1.998	1.99414	2.00781	2.002	1.9883	2.0957	1.95283
11	2.001	2.00293	1.99609	1.999	2.0059	1.95215	2.02358
12	1.9995	1.99854	2.00195	2.0005	1.9971	2.02393	1.98821
13	2.0002	2.00073	1.99902	1.9998	2.0015	1.98804	2.00590
14	1.9999	1.99963	2.00049	2.0001	1.9993	2.00598	1.99705

c. All the orbits tend to 2 in such a way that the orbit is alternately greater than 2, then less than 2.

INVESTIGATION 5: YOU CHOOSE THE SEEDS

a. The fixed point is $\frac{1}{2}$.

b. Possible answer: The table below gives the orbits for seeds of 3, 2, 0, and −5.

n	**Orbits**			
0	3	2	0	–5
1	–2	–1	1	6
2	3	2	0	–5
3	–2	–1	1	6
4	3	2	0	–5
5	–2	–1	1	6
6	3	2	0	–5
7	–2	–1	1	6
8	3	2	0	–5

c. All the orbits (except the fixed point) oscillate between two values.

INVESTIGATION 6: MATCHING ORBITS AND TIME-SERIES GRAPHS

Here are the pairings in table form:

	Orbit	**Time series**
a.	0, 2, 6, 14	E
b.	$0, -6, -8, -\frac{28}{3}$	A
c.	0, 1, 2, 3	B
d.	$0, -6, -4, -\frac{14}{3}$	C
e.	$0, -4, 0, -4$	D

INVESTIGATION 7: BE CAREFUL WITH THIS ONE!

This is a great question because it can be answered in many ways. The sure-fire way of answering it is to observe that if the orbit was created by iterating a linear function, say, $x \to ax + b$, then the following points must lie on a line:

$$(x_0, x_1), (x_1, x_2), (x_2, x_3), \ldots, (x_{n-1}, x_n)$$

If the points lie on a line, then the slope of each segment must be the same. This means that the quantity

$$\frac{x_{n+2} - x_{n+1}}{x_{n+1} - x_n}$$

must be constant because all the segments must have the same slope. To put this to use, take the orbit, create the ordered pairs defined above, and then check the slopes of each successive segment by taking differences of the *x*-values and differences of the *y*-values and then calculating their ratio. Here is how it would work for the orbit in part e:

x-differences	*x*	*y*	*y*-differences
	4	7	
3	7	13	6
6	13	25	12
12	25	49	24

Since $6/3 = 12/6 = 24/12 = 2$, the points all lie on a line with slope 2. You are now in a position to figure out exactly what the linear iteration formula is. You know it is linear and of the form $x \to 2x + b$. To find b, use the fact that you know $2(7) + b = 13$ (why?) and hence $b = -1$. Therefore, the iteration in this case is $x \to 2x - 1$.

Some of the orbits you can analyze more quickly. For example, 9, 6, 3, 0, −3 is formed by starting with 9 and taking away 3 each time or by iterating $x \to x - 3$. Using the above analysis, you will find that the orbits in parts b, d, e, and g were formed using linear iteration.

INVESTIGATION 8: PIECEWISE LINEAR ITERATION RULES

These problems can be approached in a variety of ways. Probably the easiest way is to do them by cases using the definition of the absolute value of x: $|x| = x$ if $x \geq 0$ and $|x| = -x$ if $x < 0$.

a. x is fixed for all positive x and eventually fixed at $|x|$ after one iteration if $x < 0$.

b. There are no fixed points and all orbits go to infinity.

c. 0 is fixed and all other orbits go to infinity.

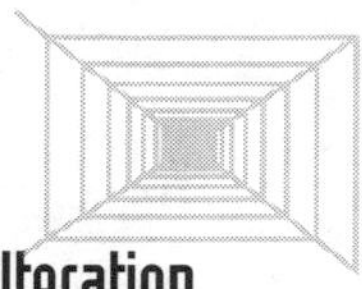

FURTHER EXPLORATION

1. **a.** The fixed point for $x \rightarrow 3x - 2$ is 1, and orbits with seeds greater than 1 go to infinity while those with seeds less than 1 go off to negative infinity.

   b. The fixed point here is -3 and all orbits are attracted to -3.

2. Suppose we have two fixed points x_0 and x_1 for the iteration rule $x \rightarrow Ax + B$. Then $Ax_0 + B = x_0$ and $Ax_1 + B = x_1$. Then we have $(A - 1)x_0 = -B$ and $(A - 1)x_1 = -B$. Therefore, $(A - 1)x_0 = (A - 1)x_1$. If $x_0 \neq x_1$, the only way this can happen is if $A - 1 = 0$ or $A = 1$. Thus, our iteration rule must be $x \rightarrow x + B$. But this iteration rule has no fixed points at all unless $B = 0$. When $B = 0$, the iteration rule is then $x \rightarrow x$, and this rule has the property that all orbits are fixed. Thus, only when $A = 1$, $B = 0$ do we have a linear iteration rule with more than one fixed point.

3.

	Condition on A	**Time series**
a.	$A = -1$	E
b.	$0 < A < 1$	B
c.	$A < -1$	D
d.	$-1 < A < 0$	C
e.	$A > 1$	A
f.	$A = 1$	G
g.	$A = 0$	F

4. The method used in Investigation 7 can be used to find the iteration rules.

	Orbit	Iteration rule
a.	$0, -3, 0, -3, \ldots$	$x \to -x - 3$
b.	$3, 3, 3, 3, \ldots$	$x \to Ax + 3(1 - A)$
c.	$2, 4, 6, 8, 10, \ldots$	$x \to x + 2$
d.	$16, -8, 4, -2, 1, \ldots$	$x \to -0.5x$
e.	$5, 3, 2, 1.5, 1.25, \ldots$	$x \to 0.5x + 0.5$
f.	$-3, 2, 2, 2, 2, \ldots$	$x \to 0x + 2$

For case b, we need to have a fixed point at 3. So we must have $3 = 3A + B$, or $B = 3(1 - A)$. Then any value of A leads to an iteration rule of the form $x \to Ax + 3(1 - A)$, which has a fixed point at 3.

5. The iteration rule $x \to 1.05x$ can be used to calculate the balances. Here they are:

Year	Balance
0	\$100.00
1	\$105.00
2	\$110.25
3	\$115.76
4	\$121.55
5	\$127.63

Year	Balance
6	\$134.01
7	\$140.71
8	\$147.75
9	\$155.13
10	\$162.89

6. The iteration rule $x \rightarrow 1.08x + 100$ can be used to calculate the balances. Here they are:

Year	Balance
0	$100.00
1	$208.00
2	$324.64
3	$450.61
4	$586.66
5	$733.59

Year	Balance
6	$892.28
7	$1,063.66
8	$1,248.76
9	$1,448.66
10	$1,664.55

7. The iteration rule $x \rightarrow 1.08x + 500$ can be used with a seed of $2000 to calculate the balances. Here they are:

Year	Balance
0	$2000.00
1	$2660.00
2	$3372.80
3	$4142.62
4	$4974.03
5	$5871.96

Continuing the table constructed above, you will see that it will take 8 years to save over $9000.

LESSON 2 ▹ TYPES OF FIXED POINTS

INVESTIGATION 1: A LINEAR ITERATION RULE

a. The fixed point is 2.

b. $4 \to 3 \to 2.5 \to 2.25 \to 2.125 \to \cdots$

$-4 \to -1 \to 0.5 \to 1.25 \to 1.625 \to \cdots$

$-1 \to 0.5 \to 1.25 \to 1.625 \to 1.8125 \to \cdots$

$6 \to 4 \to 3 \to 2.5 \to 2.25 \to 2.125 \to \cdots$

c.

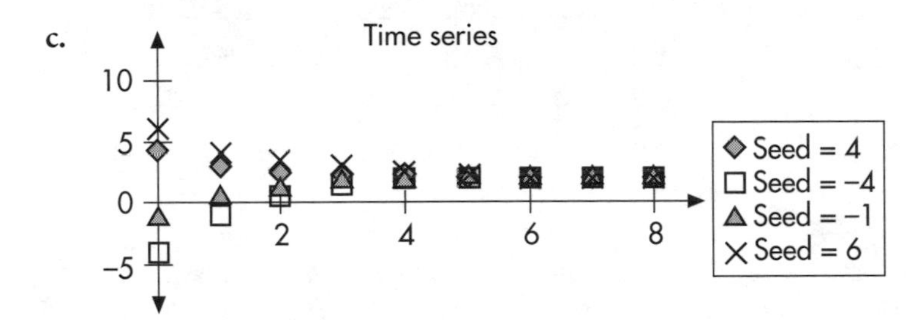

d. They all approach 2.

e. 2 is an attracting fixed point.

INVESTIGATION 2: ANOTHER LINEAR ITERATION RULE

a. The fixed point is 1.

b. $4 \to -5 \to 13 \to -23 \to 49 \to \cdots$

$-4 \to 11 \to -19 \to 41 \to -79 \to \cdots$

$-1 \to 5 \to -7 \to 17 \to -31 \to \cdots$

$0 \to 3 \to -3 \to 9 \to -15 \to \cdots$

c.

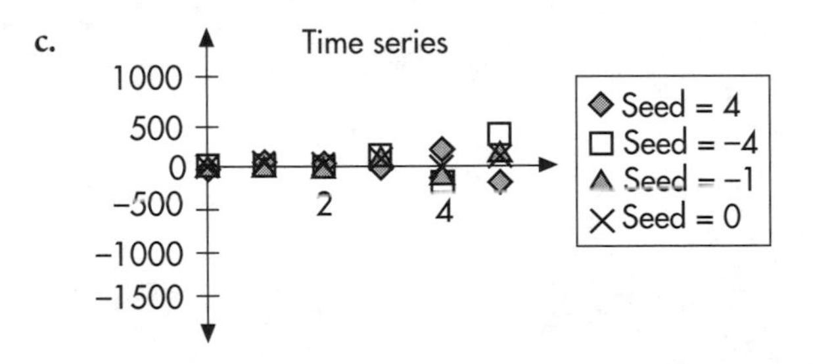

d. Except for the orbit of the fixed point, they all tend to positive or negative infinity. We can also say that the absolute value of each orbit approaches infinity.

e. 1 is a repelling fixed point.

INVESTIGATION 3: A THIRD LINEAR ITERATION RULE

a. The fixed point is –2.

b. $4 \rightarrow 7 \rightarrow 11.5 \rightarrow 18.25 \rightarrow 28.375 \rightarrow \cdots$

$-4 \rightarrow -5 \rightarrow -6.5 \rightarrow -8.75 \rightarrow -12.125 \rightarrow \cdots$

$-1 \rightarrow -0.5 \rightarrow 0.25 \rightarrow 1.375 \rightarrow 3.0625 \rightarrow \cdots$

$0 \rightarrow 1 \rightarrow 2.5 \rightarrow 4.75 \rightarrow 8.125 \rightarrow \cdots$

c.

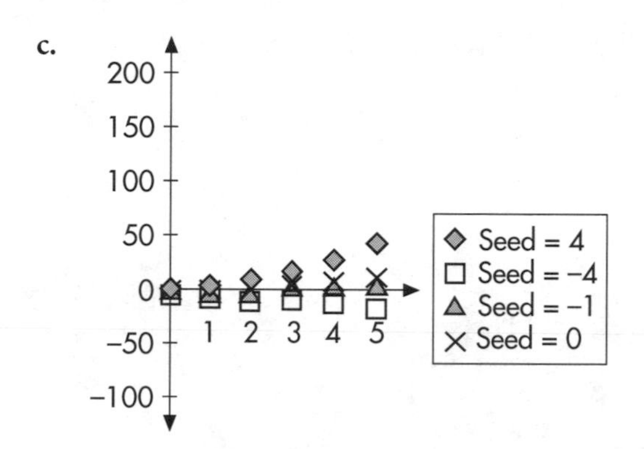

d. If the seed is greater than -2, then the orbit will take off to infinity. If the seed is less than -2, then the orbit will take off to negative infinity.

e. -2 is a repelling fixed point.

Investigation 4: One more linear iteration rule

a. The fixed point is 3.

b. $14 \rightarrow -8 \rightarrow 14 \rightarrow -8 \rightarrow 14 \rightarrow \cdots$

$-20 \rightarrow 26 \rightarrow -20 \rightarrow 26 \rightarrow -20 \rightarrow \cdots$

$-1 \rightarrow 7 \rightarrow -1 \rightarrow 7 \rightarrow -1 \rightarrow \cdots$

$0 \rightarrow 6 \rightarrow 0 \rightarrow 6 \rightarrow 0 \rightarrow \cdots$

c.

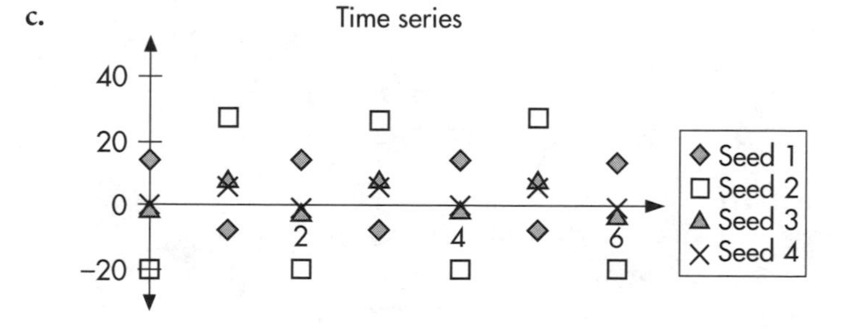

d. All orbits will be cycles of the form $a \rightarrow 6 - a \rightarrow a \rightarrow 6 - a \rightarrow \cdots$.

e. 3 is a neutral fixed point.

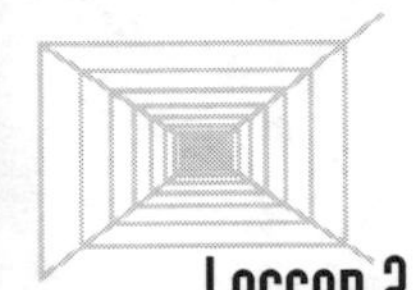

Investigation 5: And another . . .

a. The fixed point is 0.

b. $40 \to -24 \to 14.4 \to -8.64 \to 5.184 \to \cdots$

$-20 \to 12 \to -7.2 \to 4.32 \to -2.592 \to \cdots$

$-12 \to 7.2 \to -4.32 \to 2.592 \to -1.552 \to \cdots$

$6 \to -3.6 \to 2.16 \to -1.296 \to 0.776 \to \cdots$

c.

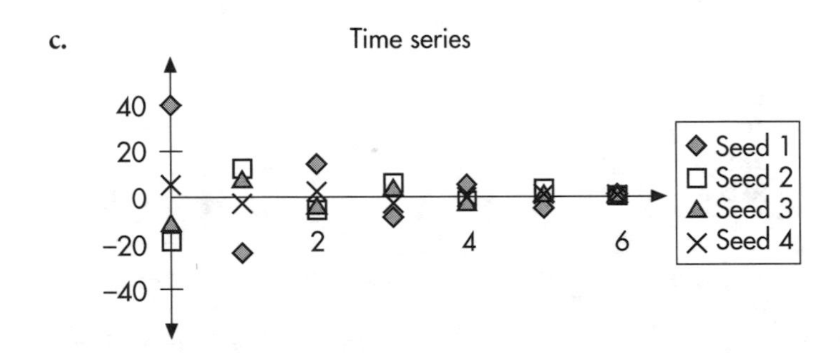

d. All orbits are attracted to 0.

e. 0 is an attracting fixed point.

Investigation 6: Determining the types of fixed points

	Iteration rule	Fixed point/type
a.	$x \to 4x + 2$	$-\frac{2}{3}$ Repelling
b.	$x \to \frac{1}{4}x + 2$	$\frac{8}{3}$ Attracting
c.	$x \to -x + 20$	10 Neutral
d.	$x \to 0.2x + 12$	15 Attracting

Investigation 7: Your fixed-point conjecture

$x \to Ax + B$ has a fixed point at $B/(1 - A)$ if $A \neq 1$. The fixed point is

- attracting if $|A| < 1$.
- repelling if $|A| > 1$.
- neutral if $A = -1$.

All points are fixed if $A = 1$, $B = 0$, and so all are neutral fixed points. There are no fixed points if $A = 1$ and B is nonzero.

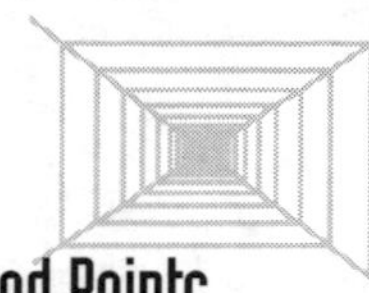

INVESTIGATION 8: A TIME-SERIES GRAPHICAL SUMMARY

Assuming that P_0 is the fixed point, then the graphs would look like the ones shown here:

a.

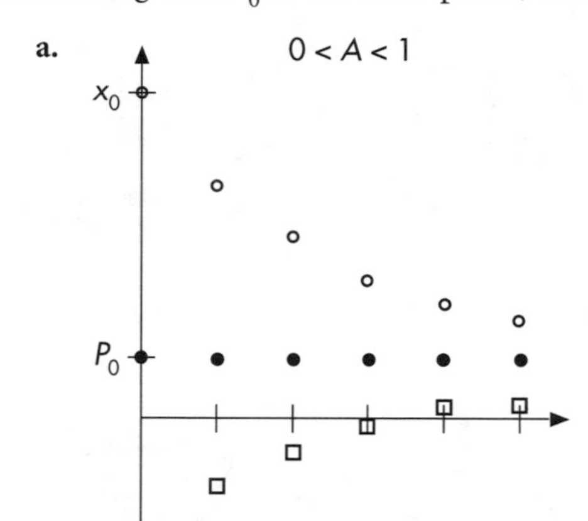

b.

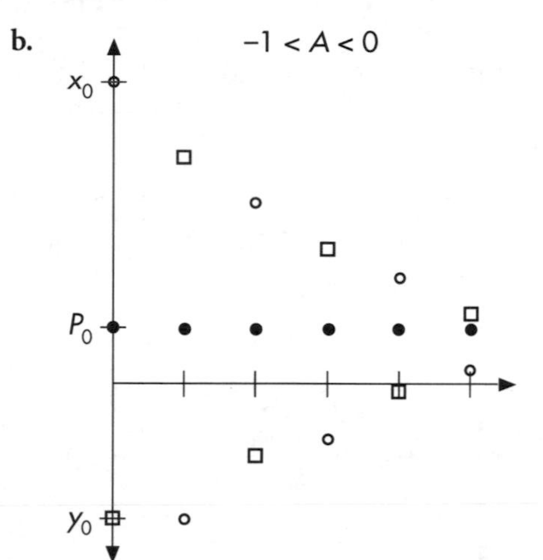

c.

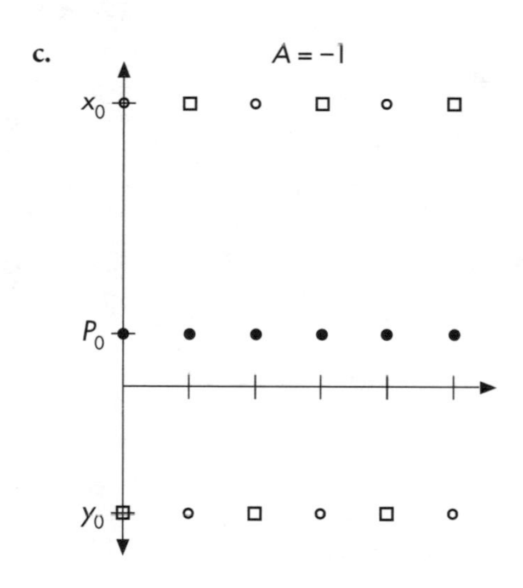

INVESTIGATION 9: FIXED POINT FREE

When $A = 1$ and $B \neq 0$, the iteration rule has no fixed points.

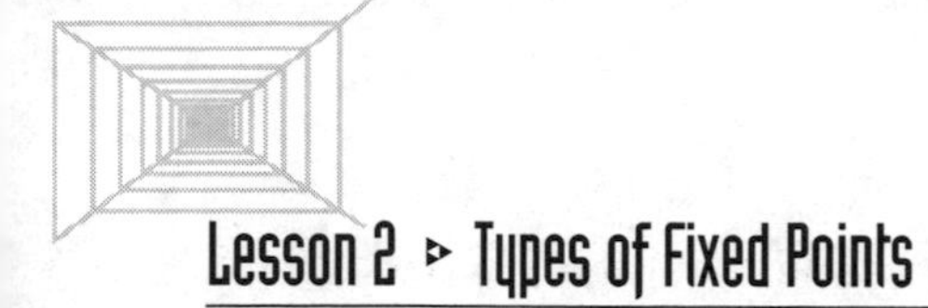

INVESTIGATION 10: MATCHING TIME-SERIES GRAPHS

	A and *B*	Time series
a.	$A = 1, B > 0$	D
b.	$A = 1, B < 0$	E
c.	$-1 < A < 0, B < 0$	A
d.	$A > 1, B > 0$	
e.	$A = 0, B > 0$	
f.	$A < -1, B > 0$	B
g.	$0 < A < 1, B < 0$	C
h.	$A > 1, B < 0$	G
i.	$A = -1, B = 0$	F

Investigation 11: The BIG picture

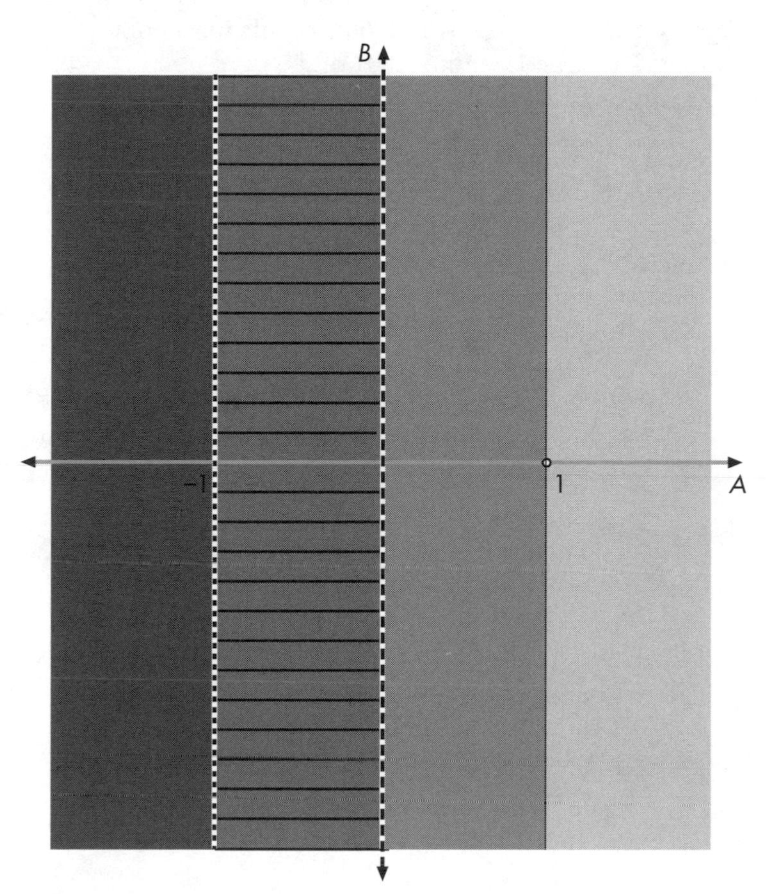

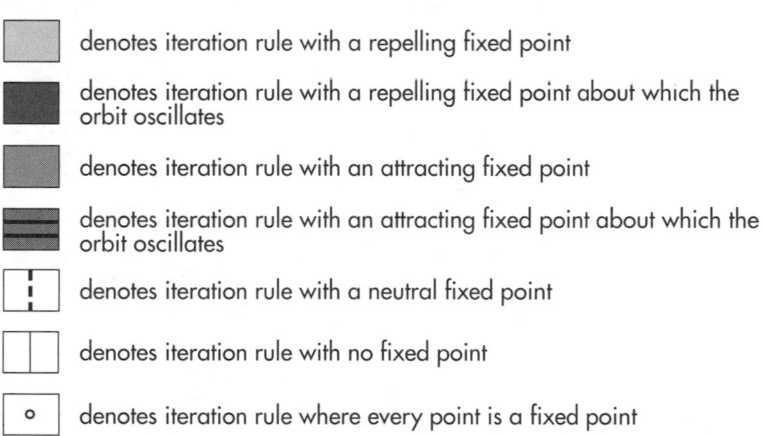

FURTHER EXPLORATION

1. The fixed point is 10. The orbit of $x_0 = 100$ is within 0.01 of this fixed point after (approximately) 88 iterations. The orbit is within 0.001 of the fixed point after (approximately) 113 iterations.

2. This can be done in several ways. One way is to start with a rule in fixed-point form such as $x \rightarrow a(x - 7) + 7$ with $-1 < a < 1$, and algebraically simplify the rule. This rule simplifies to $x \rightarrow ax - 7a + 7$. A more general way is to use the formula for the fixed point $b/(1 - a)$, set it equal to 7, and find an ordered pair that satisfies this equation. For example, if $a = -0.5$ then b would equal 10.5. The iteration rule would then be $x \rightarrow -0.5x + 10.5$, which has a fixed point of 7.

3. If the rule has a neutral fixed point and a 2-cycle, then it is of the form $x \rightarrow -x + b$. This has a fixed point at $b/2$. Since 3 is to be a fixed point, $b/2 = 3$ and $b = 6$. Therefore, the rule is $x \rightarrow -x + 6$.

4. The only possibility here is $x \rightarrow x$.

5. 0 is an attracting fixed point for $x \rightarrow 0.5|x|$.

6. 0 is a repelling fixed point for $x \rightarrow 2|x|$.

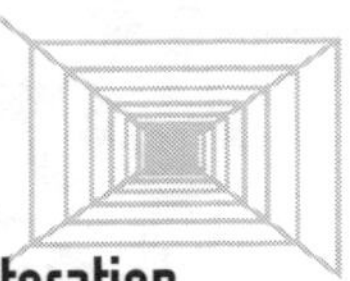

LESSON 3 ▹ GRAPHICAL ITERATION

INVESTIGATION 1: GRAPHICAL ITERATION AND TIME SERIES

a. $x_0 = 0$

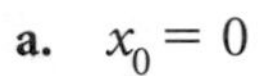

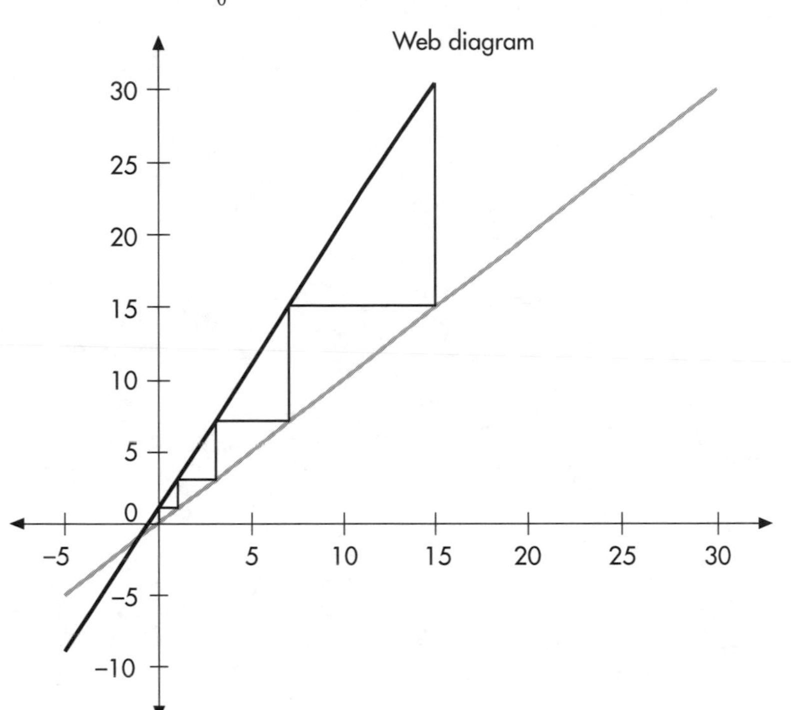

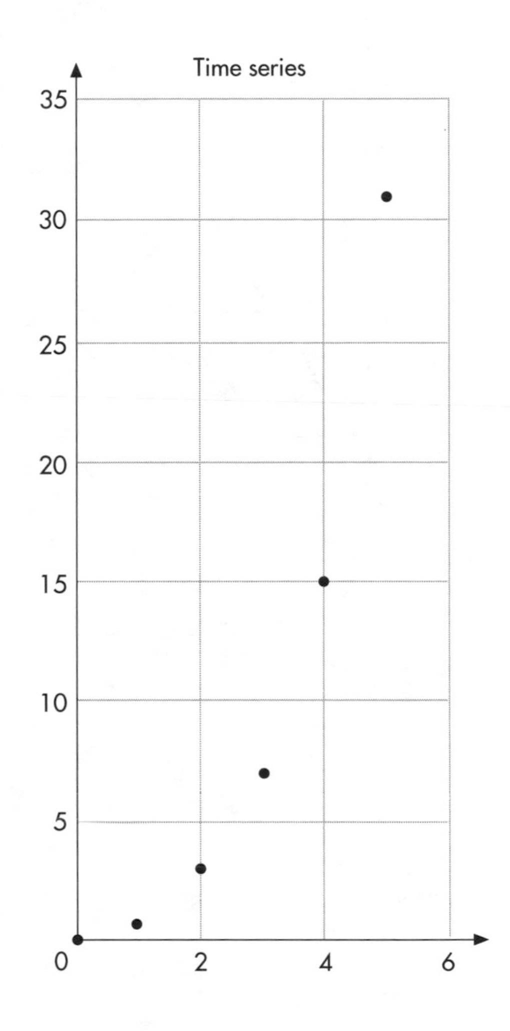

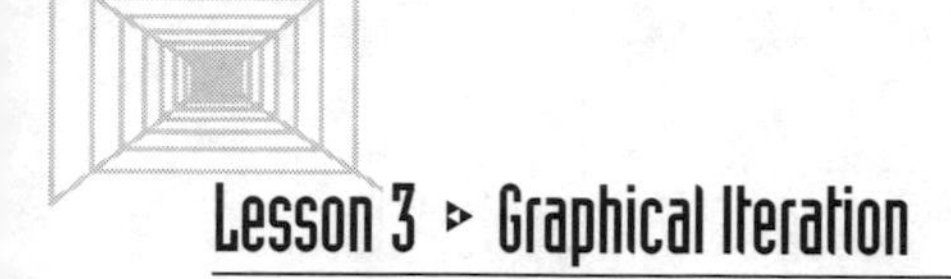

b. $x_0 = 1.5$

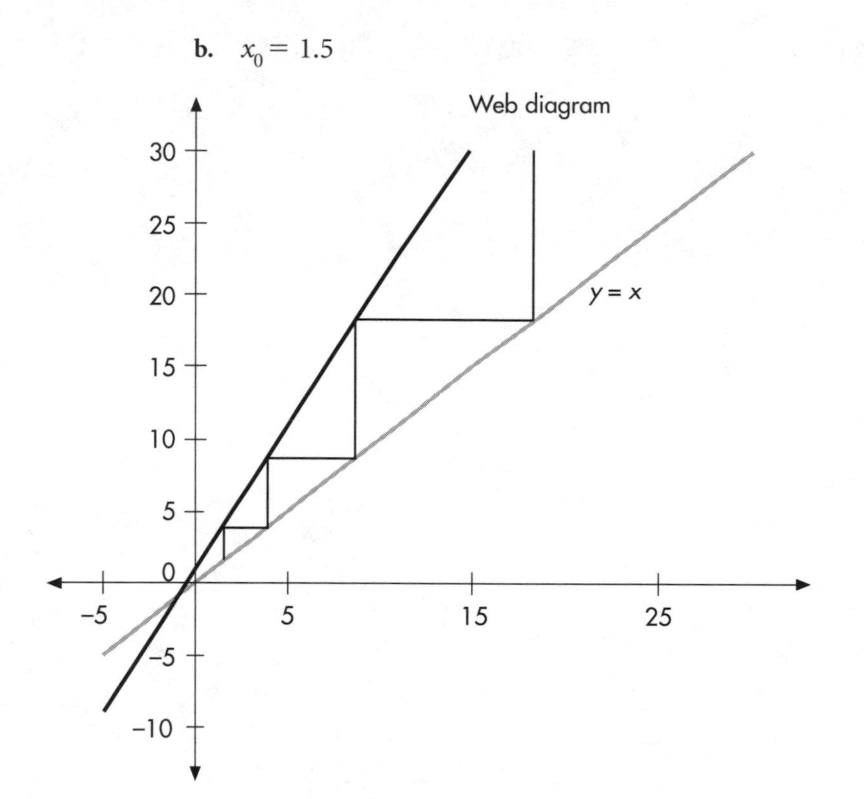
Web diagram
30
25
20
15
10
5
0
-5
-10
-5
5
15
25
y = x

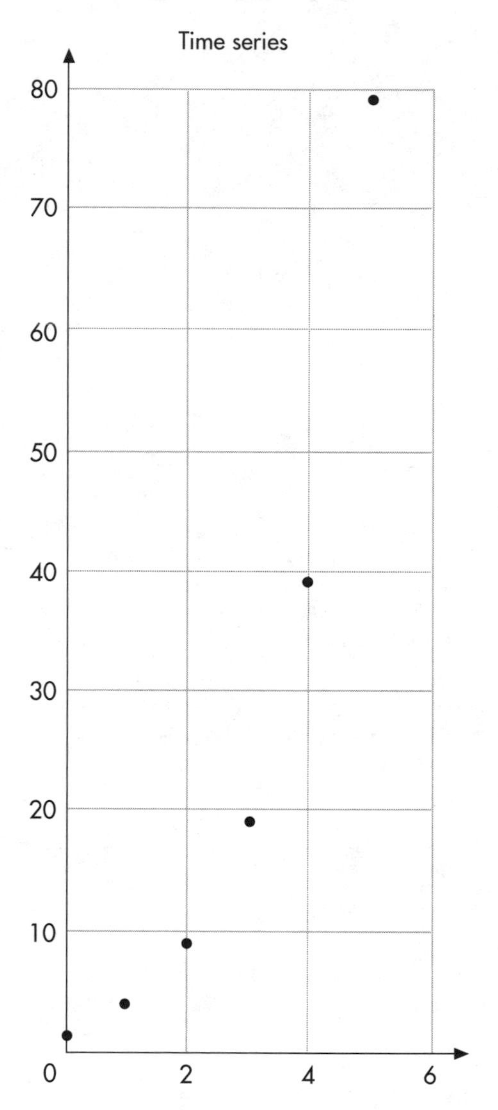
Time series
80
70
60
50
40
30
20
10
0
2
4
6

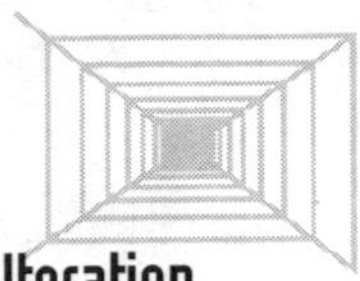

c. $x_0 = -1.5$

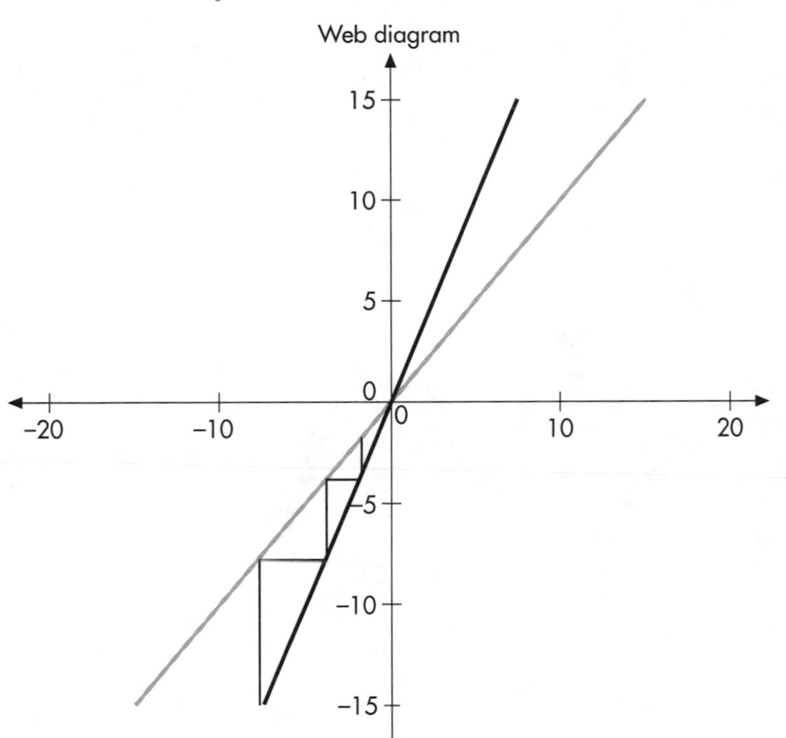
Web diagram
15
10
5
0
−5
−10
−15
−20
−10
0
10
20

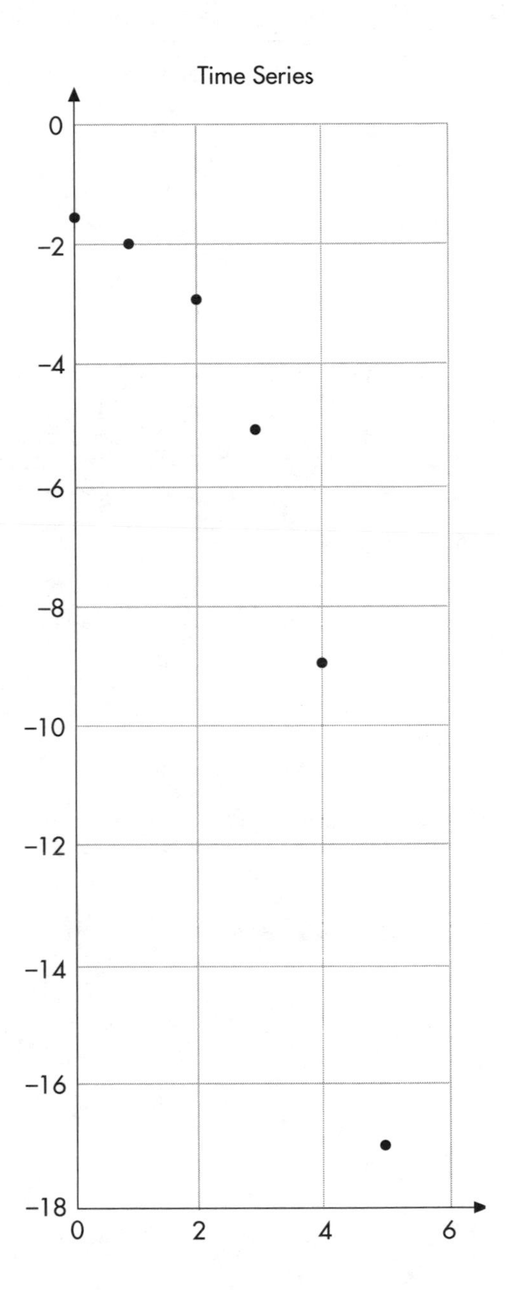
Time Series
0
−2
−4
−6
−8
−10
−12
−14
−16
−18
0
2
4
6

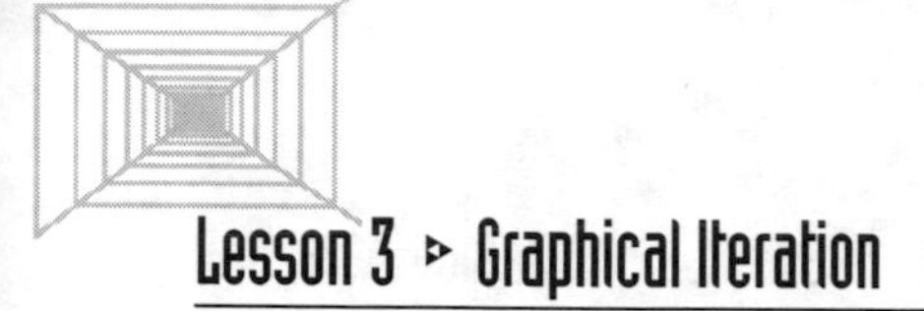

INVESTIGATION 2: FINDING THE FATE OF THE ORBIT GEOMETRICALLY

a. The orbit shown is attracted to the fixed point $\frac{40}{17}$.

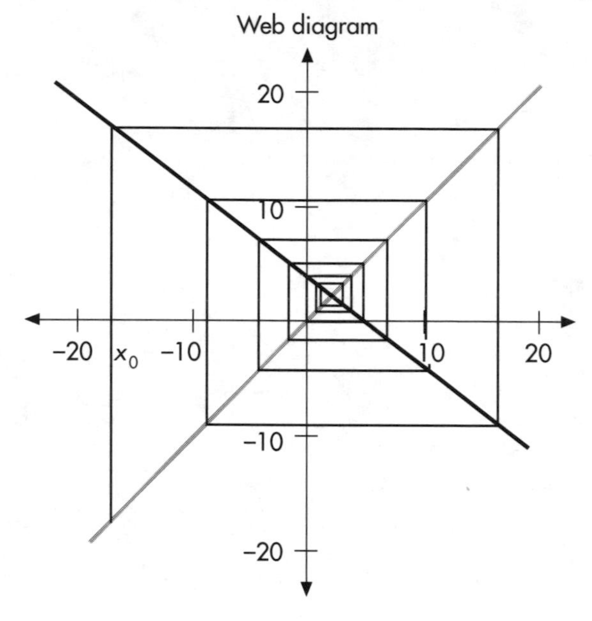

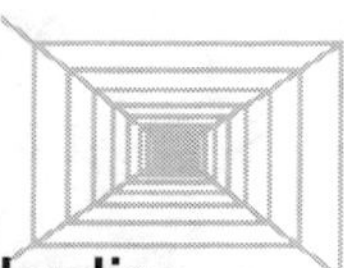

b. The orbit shown is attracted to the fixed point $\frac{40}{3}$.

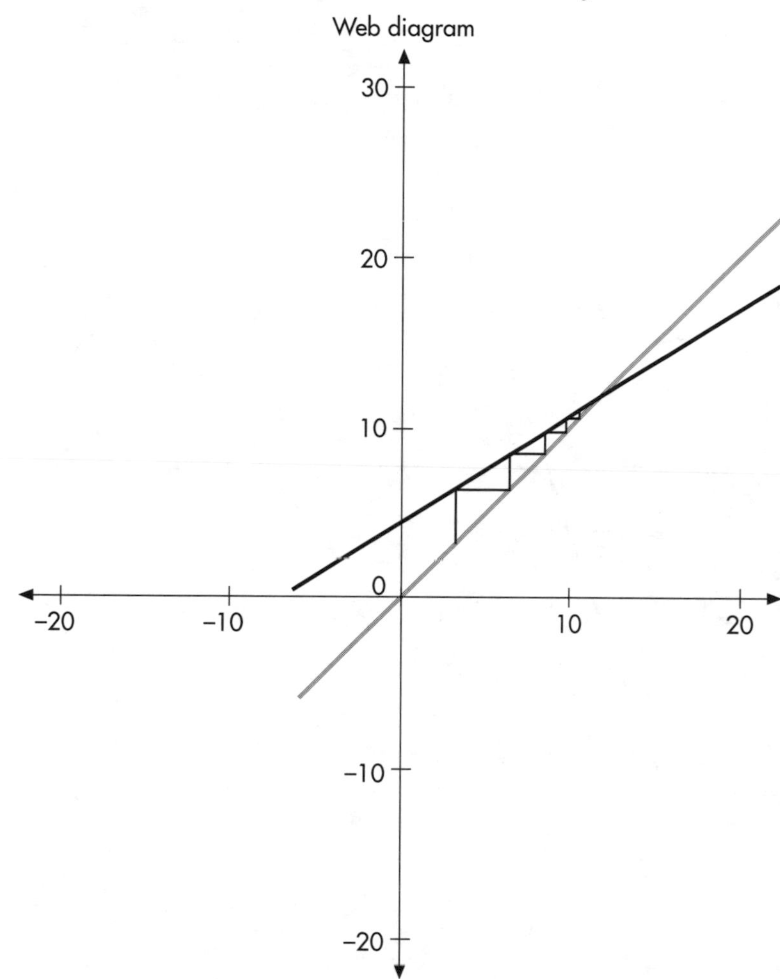

INVESTIGATION 3: GRAPHICAL ITERATION

a. $x \rightarrow 3x - 1, x_0 = 0$

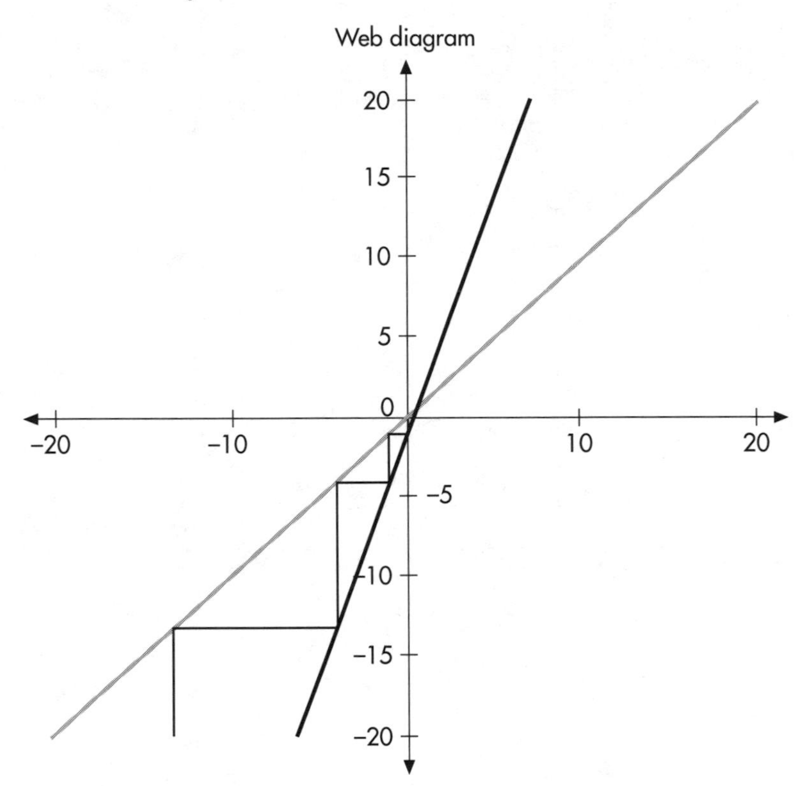

b. $x \to \frac{1}{2}x + 2$, $x_0 = -20$

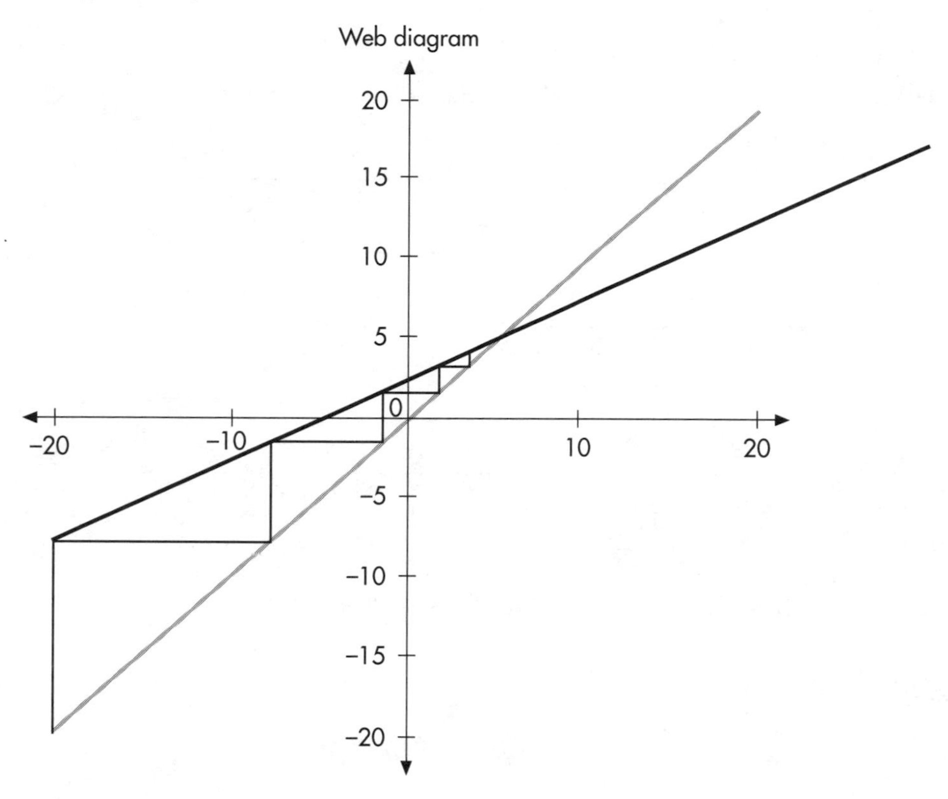

c. $x \to -3x - 1$, $x_0 = 0$

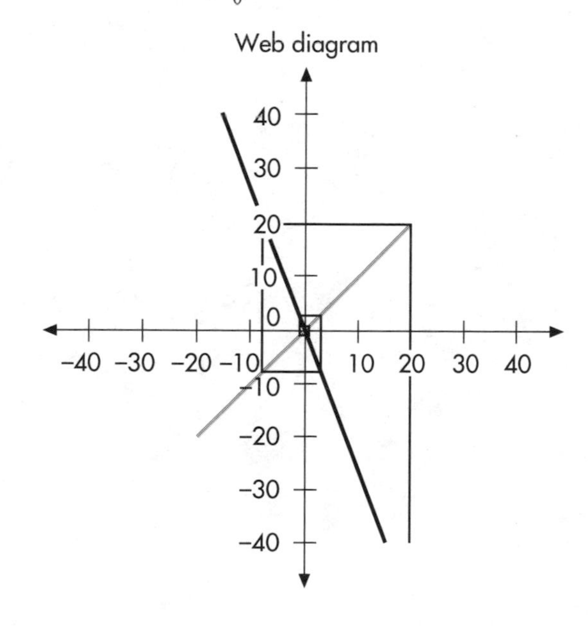

INVESTIGATION 4: MATCHING ORBITS, TIME SERIES, AND GRAPHICAL ITERATION

	Orbit	Time series	Graphical iteration	$x \to Ax + B$
a.	0, 2, 6, 14	E	III	$x \to 2x + 2$
b.	$0, -6, -8, -\frac{26}{3}$	A	V	$x \to \frac{1}{3}x - 6$
c.	0, 1, 2, 3	B	II	$x \to x + 1$
d.	$0, -6, -4, -\frac{14}{3}$	C	IV	$x \to -\frac{1}{3}x - 6$
e.	$0, -4, 0, -4$	D	I	$x \to -x - 4$
f.	0, 3, 3, 3	F	VI	$x \to 3$

INVESTIGATION 5: GEOMETRIC MATCHING

Web graph	Time series
1	A
2	G
3	E
4	B
5	F
6	C

INVESTIGATION 6: NEUTRAL FIXED POINTS

a.

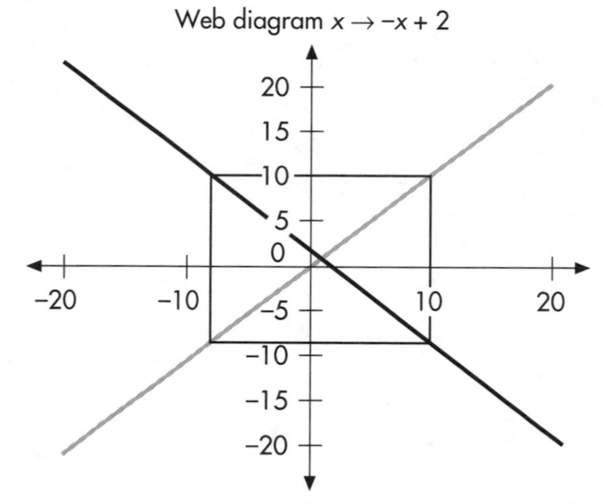

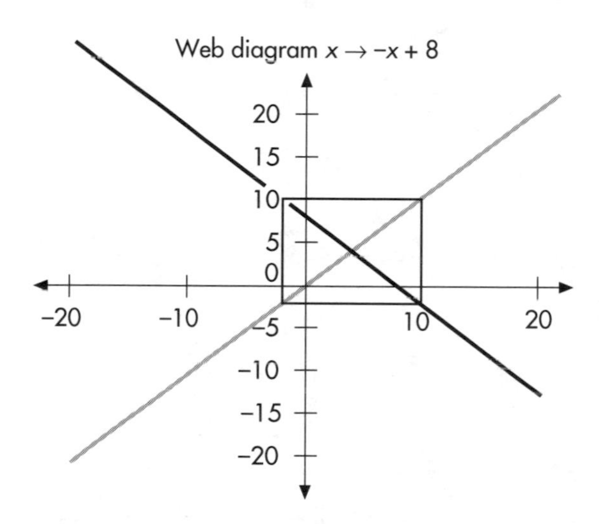

In the two graphs above, the seed is 10 and both graphs were iterated by iterating a rule of the form $x \rightarrow -x_0 + B$. In the first case, $B = 2$ and the orbit generated was $10, -8, 10, -8, \ldots$ and in the second case, $B = 8$ and the orbit was $10, -2, 10, -2, \ldots$. In general, the orbits will all be of the form $x_0, -x_0 + B, x_0, -x_0 + B, \ldots$.

b. The orbits are all 2-cycles of the form $x_0, -x_0 + B, x_0, -x_0 + B, \ldots$.

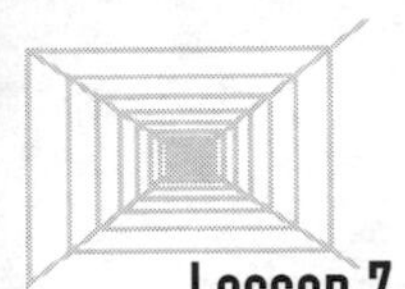

c.

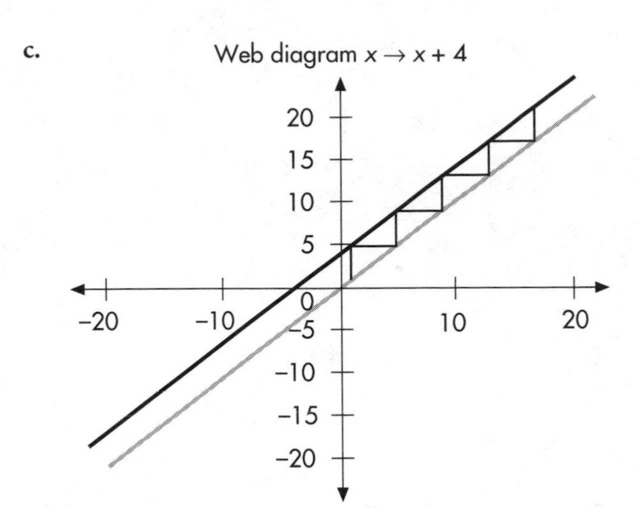

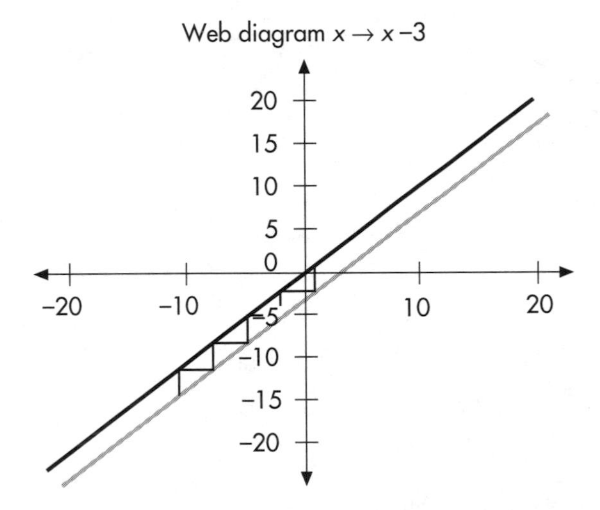

d. The graphs above show what happens in the case $x \rightarrow x + B$ when $B > 0$ and the case when $B < 0$. In the first case the orbit goes off to infinity, and in the second the orbit goes off to negative infinity. Notice that in both cases there is no fixed point. This shouldn't be a surprise because the two lines are parallel and $B/(1 - A)$ is undefined.

INVESTIGATION 7: BACK TO THE PARAMETER PLANE

A	B	Orbit
$A > 1$	all values	Orbits are repelled from the fixed point F. If $x_0 > F$, orbit will go off to infinity. If $x_0 < F$, orbit will go off to negative infinity.
$A = 1$	$B > 0$ $B = 0$ $B < 0$	Orbits go to infinity. All points fixed. Orbits go to negative infinity.
$0 < A < 1$		Orbits are attracted to the fixed point F. If $x_0 > F$, F is approached from above. If $x_0 < F$, F is approached from below.
$A = 0$	all values	B is a fixed point for all x_0.
$-1 < A < 0$	all values	Orbits are attracted to the fixed point F. If $x_0 > F$, orbit will oscillate: first to the left of F, then to the right. If $x_0 < F$, orbit will oscillate: first to the right of F, then to the left.
$A = -1$	all values	The fixed point F is neutral, and the orbits will oscillate between x_0 and $-x_0 + B$.
$A < -1$	all values	Orbits are repelled from the fixed point and will get larger in absolute value and take off alternately to positive and negative infinity.

INVESTIGATION 8: BACK TO THE PARAMETER PLANE

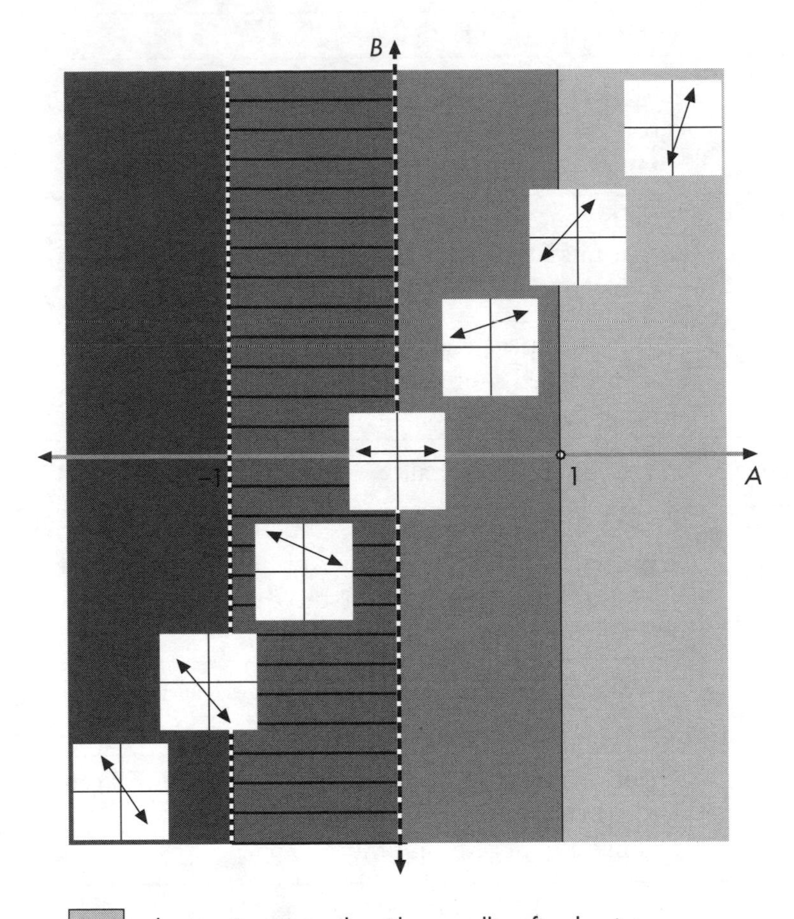

denotes iteration rule with a repelling fixed point

denotes iteration rule with a repelling fixed point about which the orbit oscillates

denotes iteration rule with an attracting fixed point

denotes iteration rule with an attracting fixed point about which the orbit oscillates

denotes iteration rule with a neutral fixed point

denotes iteration rule with no fixed point

denotes iteration rule where every point is a fixed point

denotes iteration rule where every point is a fixed point after one iteration

INVESTIGATION 9: ORBITS FOR LINEAR ITERATION RULES

	Orbit	Iteration rule
a.	0, 2, 1, 0, 1, 2, . . .	not linear
b.	8, 4, 2, 1, . . .	$x \rightarrow 0.5x$
c.	−3, −3, −3, −3, . . .	$x \rightarrow A(x + 3) - 3$
d.	6, 4, 2, 0, . . .	$x \rightarrow x - 2$
e.	6, 4, 2, 10, . . .	not linear
f.	−3, 5, −3, 5, . . .	$x \rightarrow -x + 2$

FURTHER EXPLORATION

1.

	Iteration rule	Fixed point/type
a.	$x \rightarrow -7.2x - 1$	$-\frac{1}{8.2}$ Repelling
b.	$x \rightarrow -0.72x - 1$	$\frac{1}{1.72}$ Attracting
c.	$x \rightarrow -720x - 1$	$-\frac{1}{721}$ Repelling
d.	$x \rightarrow -\frac{1}{72}x - 1$	$-\frac{72}{73}$ Attracting
e.	$x \rightarrow -\frac{1}{72}x - x$	0 Repelling
f.	$x \rightarrow 7x - 72$	12 Repelling

2.

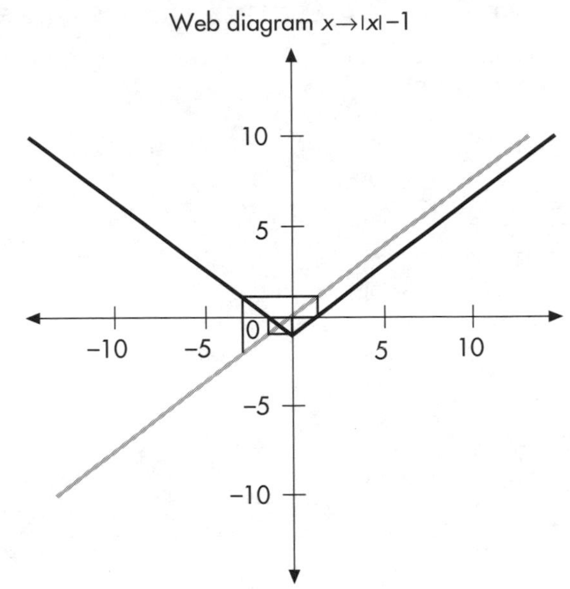

This is an interesting problem. First of all, the fixed point is $-\frac{1}{2}$, which can be found by finding the intersection of $y = x$ and $y = |x| - 1$. The point is neutral since the slope of the graph is -1 for $x < 0$ and 1 for $x > 0$.

3. Here is a table with the orbits. Other interesting seeds to try are

- $\frac{1}{p}$ where p is a prime number,
- $\frac{1}{a}$ where a is a power of 2, and
- $\frac{1}{k}$ where k is a number of the form $2^n - 1$ where n is a positive integer.

	Seed	Orbit
a.	$\frac{1}{3}$	$\frac{1}{3} \to \frac{2}{3} \to \frac{1}{3} \to \frac{2}{3} \to \cdots$
b.	$\frac{1}{7}$	$\frac{1}{7} \to \frac{2}{7} \to \frac{4}{7} \to \frac{1}{7} \to \cdots$
c.	$\frac{1}{8}$	$\frac{1}{8} \to \frac{1}{4} \to \frac{1}{2} \to 0 \to 0 \to \cdots$
d.	$\frac{3}{7}$	$\frac{3}{7} \to \frac{6}{7} \to \frac{5}{7} \to \frac{3}{7} \to \cdots$
e.	$\frac{1}{9}$	$\frac{1}{9} \to \frac{2}{9} \to \frac{4}{9} \to \frac{8}{9} \to \frac{7}{9} \to \frac{5}{9} \to \frac{1}{9} \to \cdots$

a. The orbit of $\frac{1}{3}$:

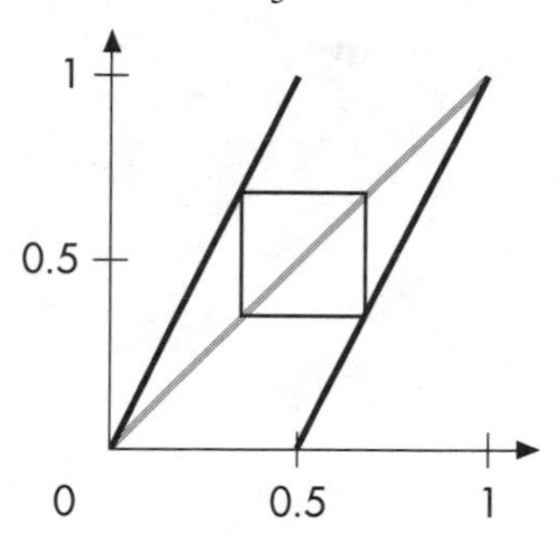

b. The orbit of $\frac{1}{7}$:

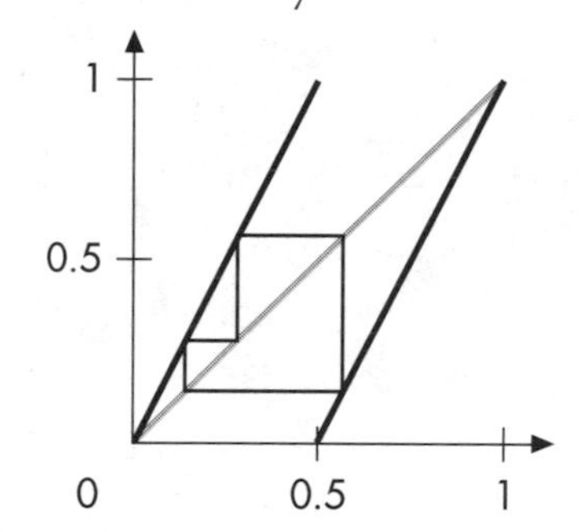

c. The orbit of $\frac{1}{8}$:

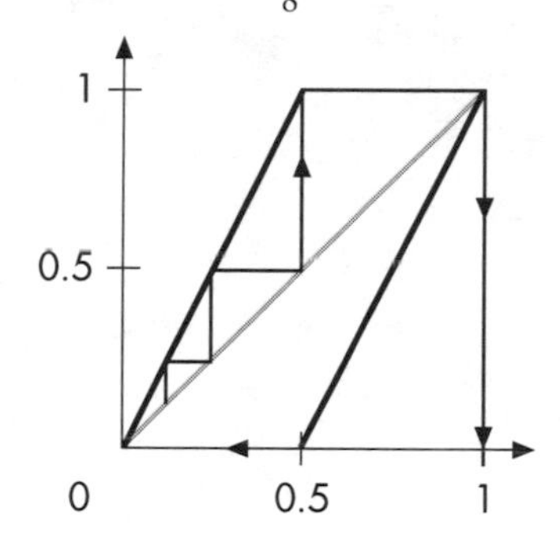

d. The orbit of $\frac{3}{7}$:

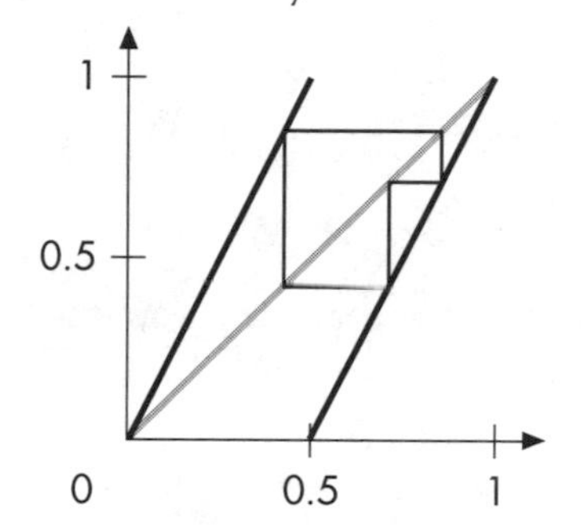

e. The orbit of $\frac{1}{9}$:

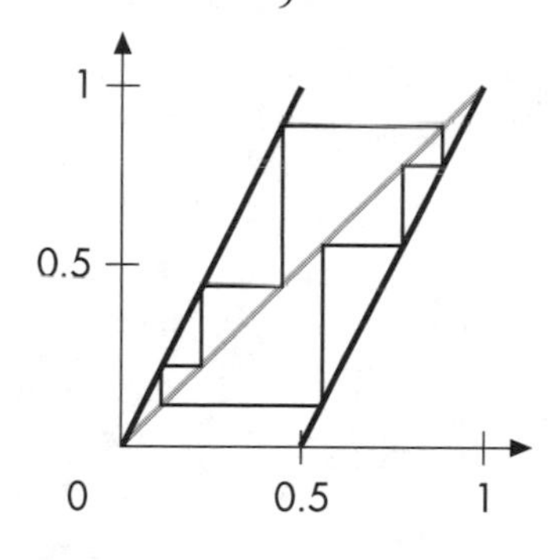

Lesson 4 ▹ Measuring Population Growth

Investigation 1: The exponential growth model

a. The population will become extinct for $0 < k < 1$.

b. The population will increase without bound for $k > 1$.

INVESTIGATION 2: IMPROVING THE MODEL

Some of the factors the exponential growth model does not take into consideration might include

- the carrying capacity of the environment,
- randomly occurring natural disasters,
- randomly occurring epidemics, or
- immigration and emigration.

INVESTIGATION 3: FINDING THE FIXED POINTS

To find the fixed points, you must solve $x = kx(1 - x)$. The solutions are $x_1 = 0$ and $x_2 = (k - 1)/k$. For both fixed points to be non-negative, k must be greater than 1.

INVESTIGATION 4: AN EXPERIMENT

a. All the orbits go to 0.

b. This means that the population is dying off.

c. The same thing happens for any k-value between 0 and 1.

d. 0 is a fixed point for any value of k.

e. When $k = 1$, the population ever so slowly dies off. The closer the seed is to 1 the longer it takes. If k is slightly greater than 1, say 1.1, then there is a fixed point that can be found at $x = (k - 1)/k$ as shown in Investigation 3. If $k = 1.1$, then $x = 1/11$ is a fixed point.

INVESTIGATION 5: ANOTHER EXPERIMENT

a.

Value of k	Fate of orbit
1.3	approaches fixed point at 0.2307. . .
1.8	approaches fixed point at $\frac{4}{9}$
2	approaches fixed point at 0.5
2.5	approaches fixed point at 0.6

b. Population will stabilize after a period of time.

c. For all values of k that are less than 3, the orbit approaches or actually ends up equal to the fixed point.

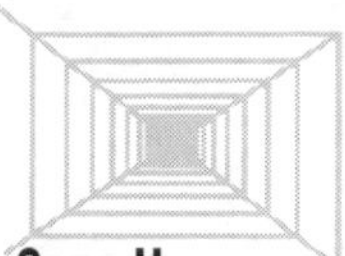

d. When $k = 3$, the orbit appears to oscillate between two values. Actually, this orbit approaches a fixed point, but it takes many thousands of iterations to get there. When k is greater than 3, it clearly cycles between two values. If $k = 3.1$, the orbit appears to cycle between 0.558 . . . and 0.764 . . . .

INVESTIGATION 6: LARGER *k*-VALUES

a.

Value of *k*	Attracting cycle
3.2	tends to a 2-cycle: 0.5130, 0.7995
3.4	tends to a 2-cycle: 0.452, 0.842
3.5	tends to a 4-cycle: 0.5009, 0.8749, 0.3828, 0.8269
3.554	tends to an 8-cycle: 0.5007, 0.8884, 0.3520, 0.8107, 0.5453, 0.8812, 0.3720, 0.8303

b. This means that as k increases from 3 to 3.554, the population will go from stabilizing at a single value, to oscillating between two values (a 2-cycle), to oscillating between 4 values (a 4-cycle), to oscillating between 8 values (an 8-cycle). Note that as k increases through the given range the orbits keep splitting in two.

c. It should come as no surprise that if you let k get slightly larger than 3.554, you will be able to produce a 16-cycle. When a cycle splits into a new cycle with twice as many points, the orbit is said to bifurcate.

INVESTIGATION 7: THE ORBIT DIAGRAM

c.

Value of *k*	Behavior of orbit
$0 < k < 1$	goes to 0
$1 < k < 2$	goes to a fixed point
$2 < k < 3$	goes to a fixed point
$3 < k < 3.44$	attracted to a 2-cycle
$3.46 < k < 3.541$	attracted to a 4-cycle
$3.543 < k < 3.563$	attracted to an 8-cycle
$3.565 < k < 3.568$	attracted to a 16-cycle

Here is a computer-generated picture of what really results from this process:

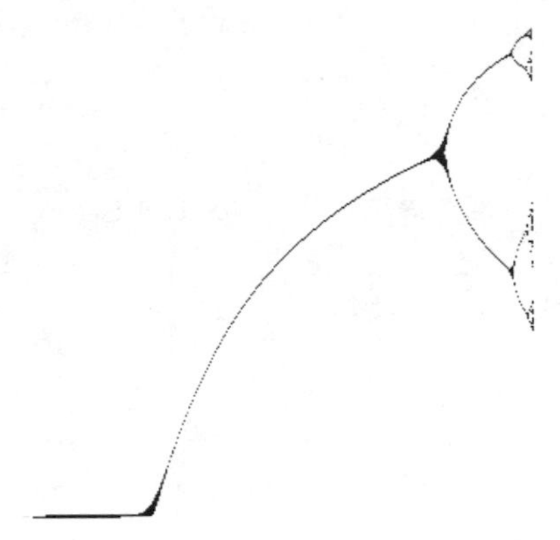

e. As k approaches 3.57, the orbit begins to approach a 32-cycle. You will be able to see this if you generate at least 500 iterations.

f. What happens here is neat. For values of k slightly less than 3.83, the orbits are chaotic. But with a slight change in the k-value, the orbits suddenly settle into a 3-cycle and then start to bifurcate going from cycles of period 3 to 6 to 12 to 24, and so on.

Further Exploration

1. This exploration is also carried out in Investigations throughout Lesson 10. Answers are recorded there.

   Quickly, the orbit of 0 goes to a fixed point for $-0.75 < c \leq 0.25$; a 2-cycle for $-1.25 < c \leq -0.75$; then a 4-cycle, 8-cycle, 16-cycle, ...

Lesson 5 ▸ Nonlinear Iteration

Investigation 1: Graphical iteration

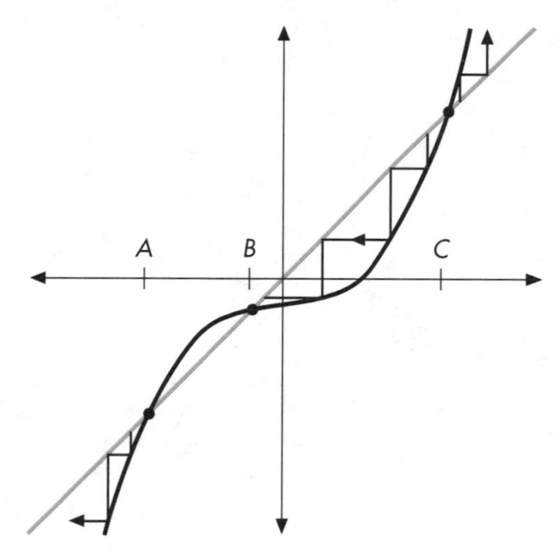

	Seed	Fate of orbit
a.	$x > C$	goes off to infinity
b.	$B < x < C$	attracted to B
c.	$A < x < B$	attracted to B
d.	$x < A$	goes off to negative infinity

e. A and C are repelling and B is attracting.

Investigation 2: The logistic iteration rule

a. 0 is a repelling fixed point and 0.5 is an attracting fixed point.

b. 0 is a repelling fixed point and 0.6 is an attracting fixed point.

c. 0 and $\frac{11}{16}$ are both repelling fixed points.

d. The orbits in both parts a and b are attracted to fixed points. The orbit in part c appears to cycle between two values.

Investigation 3: Other Iteration Rules

a. The fixed point at 0 is repelling.

Web diagram for $x \to 2x - x^2$

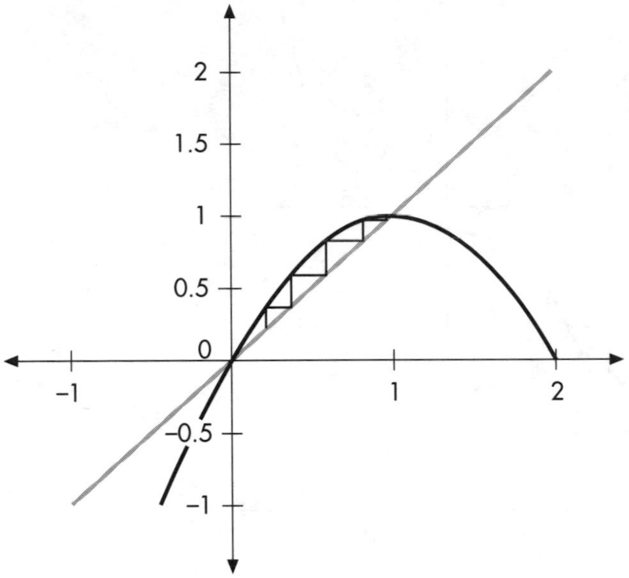

b. The fixed point at 0 is attracting.

Web diagram for $x \to x^2 + \frac{x}{2}$

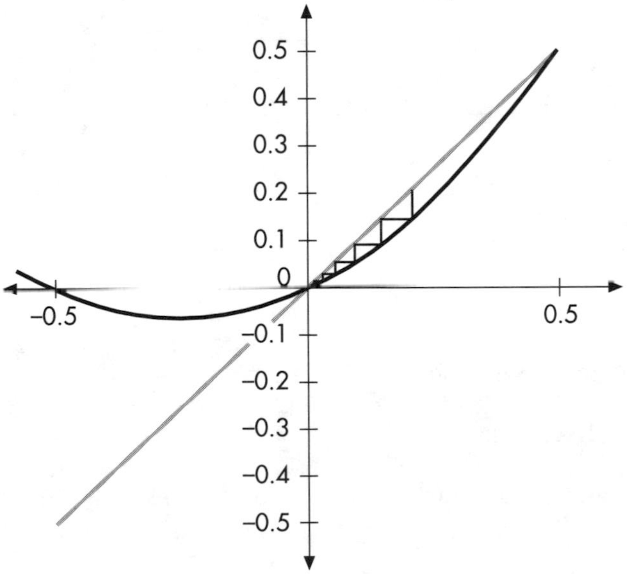

c. The fixed point at 0 is attracting.

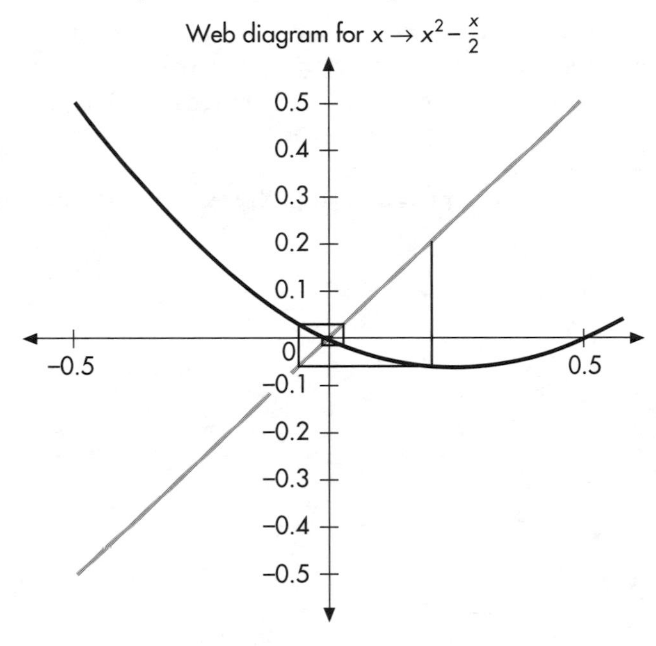

d. The fixed point at 0 is repelling.

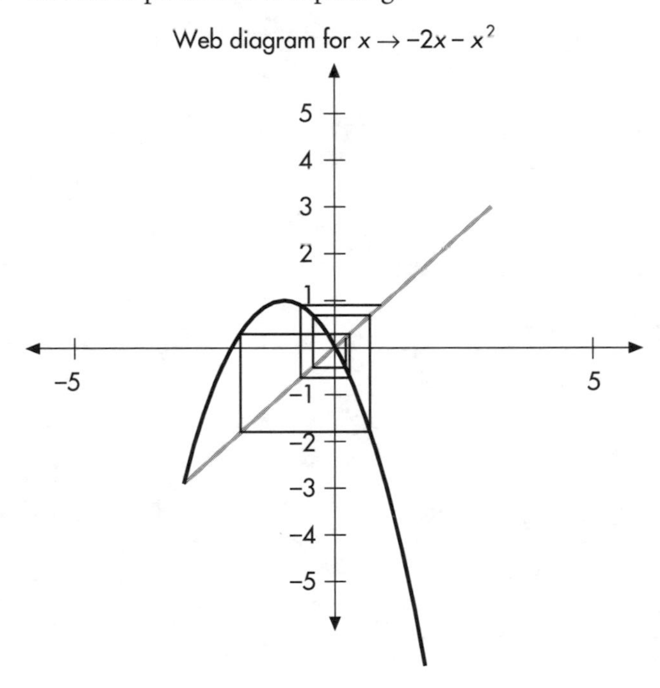

INVESTIGATION 4: DETERMINING THE FATE OF ORBITS GRAPHICALLY

Since the iteration function is defined only for $x \geq 0$, we need only look at seeds greater than or equal to 0. There are two fixed points: a repelling one at $x = 0$ and an attracting one at $x = 1$. With the exception of 0 and 1, which are fixed, all other seeds produce orbits that are attracted to 1.

INVESTIGATION 5: GRAPHICAL ITERATION AND TIME-SERIES GRAPHS

a.

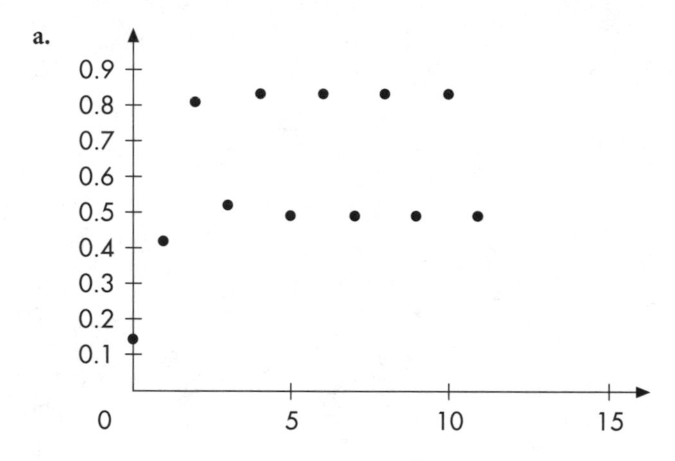

b.

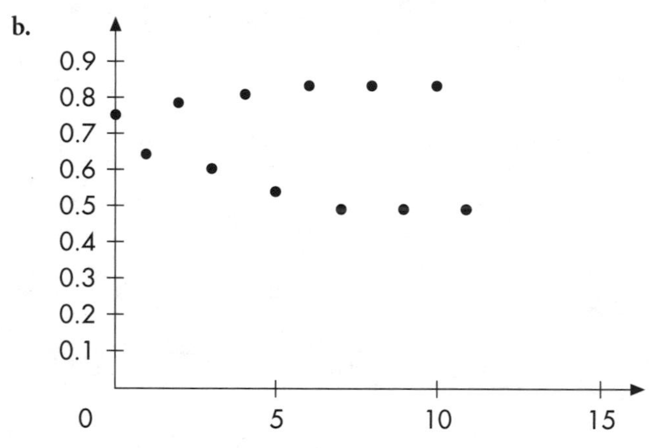

c.

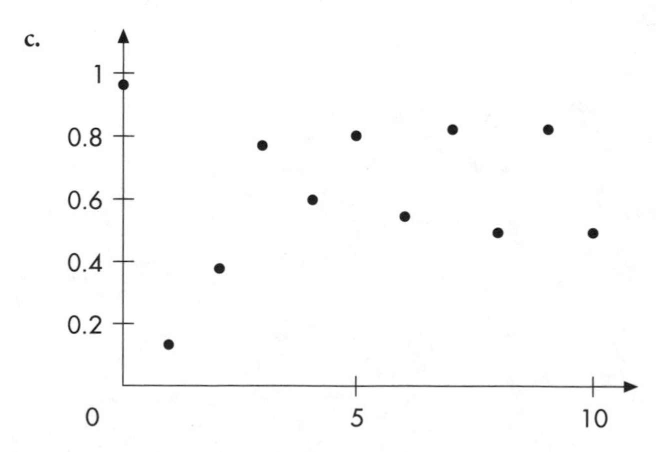

INVESTIGATION 6: MATCHING GRAPHS

Time series	Web diagram
A	VI
B	IV
C	II
D	III
E	V
F	I

INVESTIGATION 7: TARGET PRACTICE

Here are the seeds that will produce the desired sequence of A's and B's.

a. To create AABB, use $x_0 = 0.05$.

b. To create ABBAB, use $x_0 = 0.2$.

c. To create BBBBA, use $x_0 = 0.7$.

a.

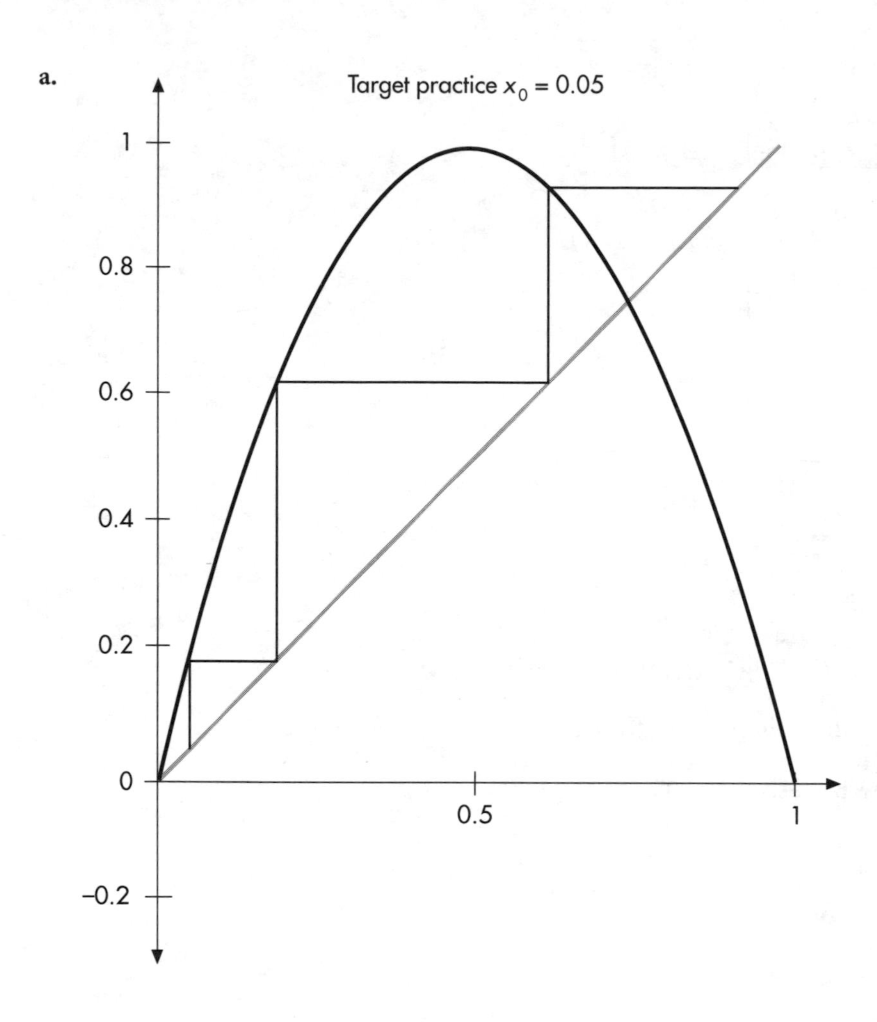

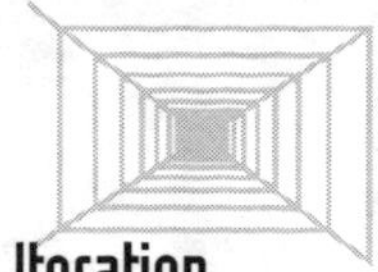

b.

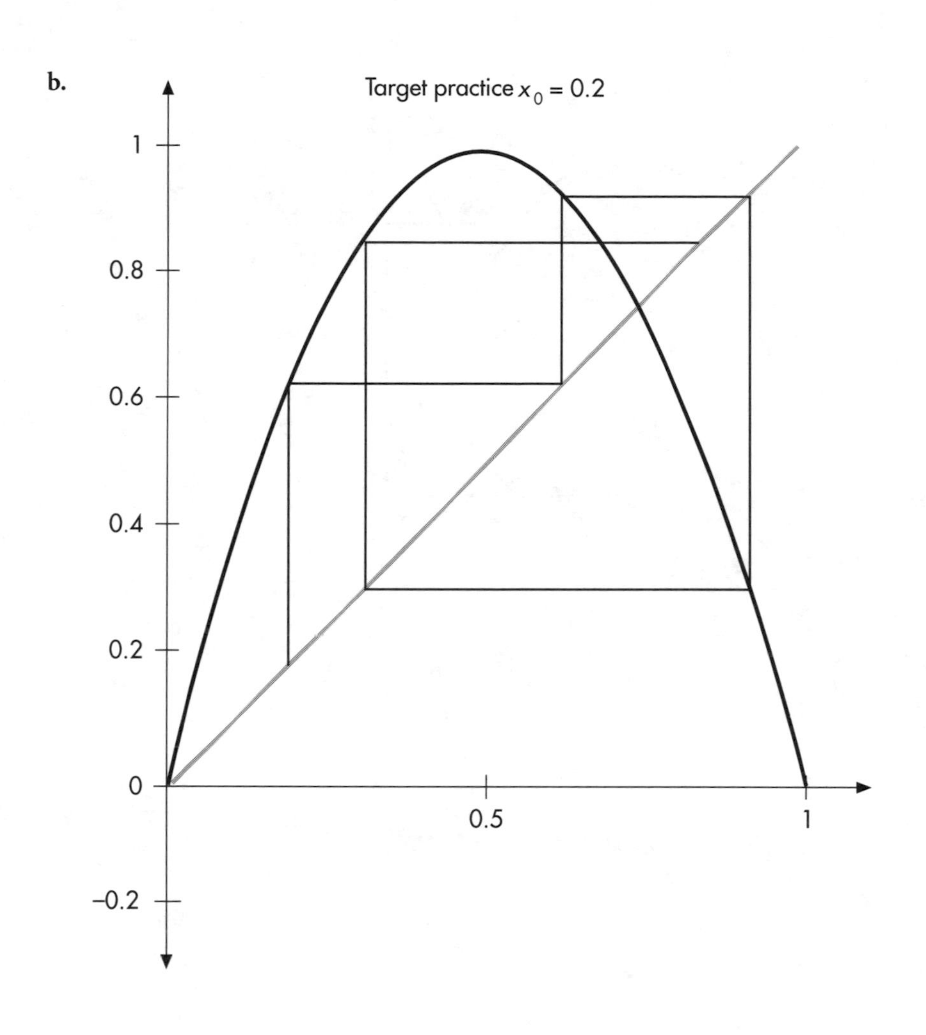

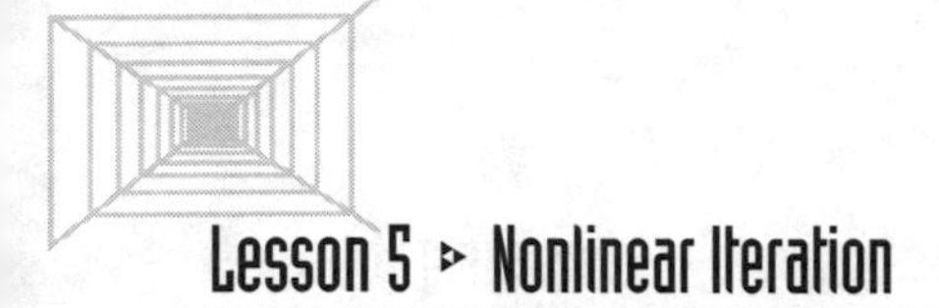

c.

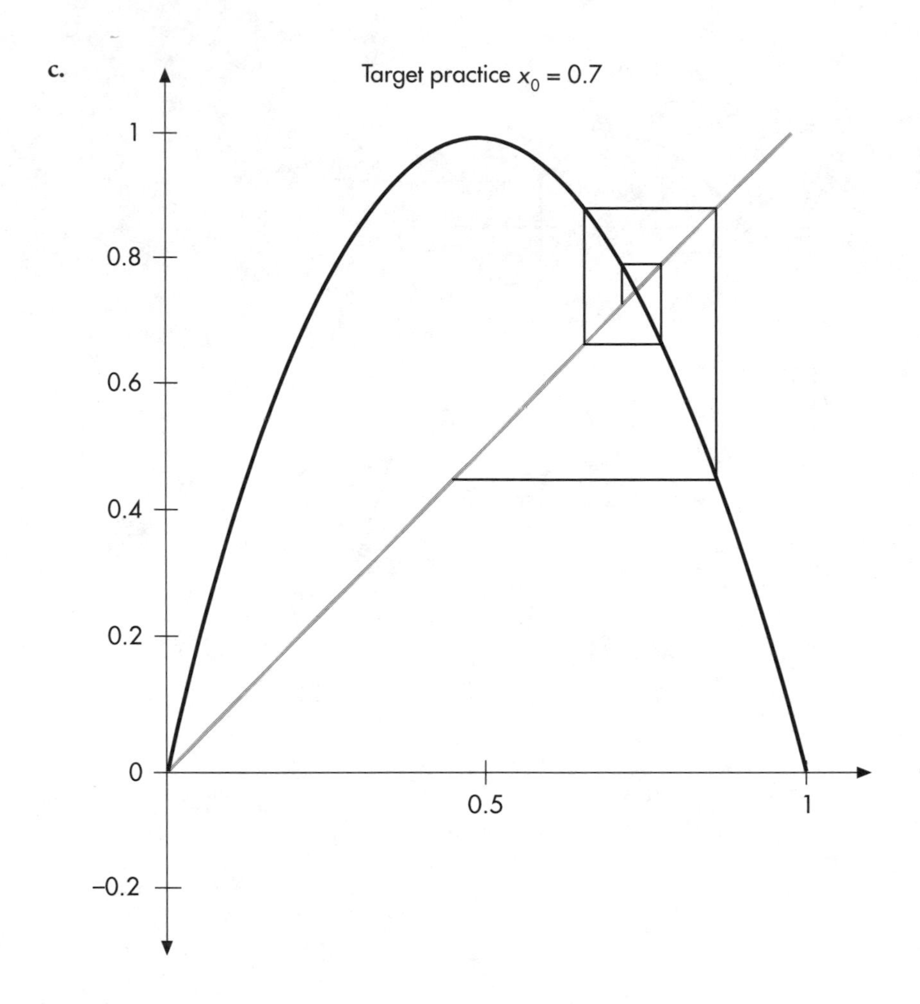

INVESTIGATION 8: CYCLE PRACTICE

These problems are tough. It is impossible to find seeds that yield cycles exactly. The idea is to come up with seeds whose orbits "close up" at least approximately after the required voyage through A's and B's. Here are some:

a.

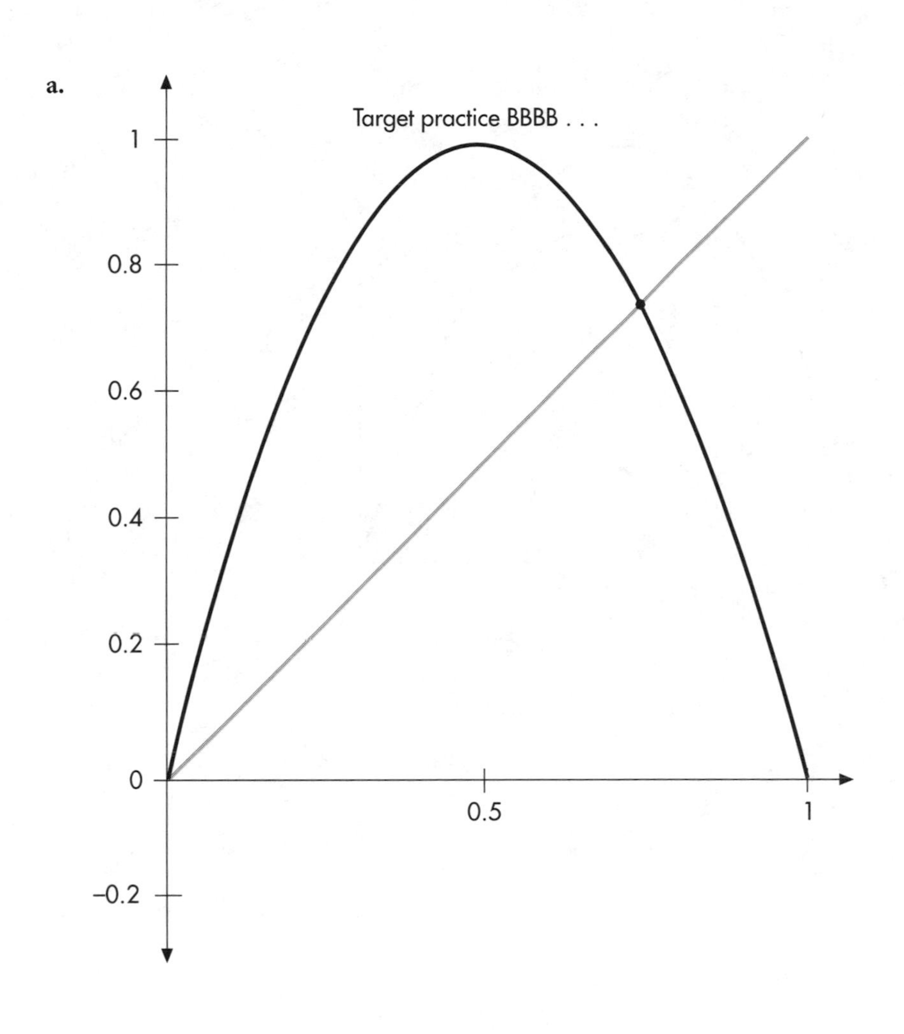

b.

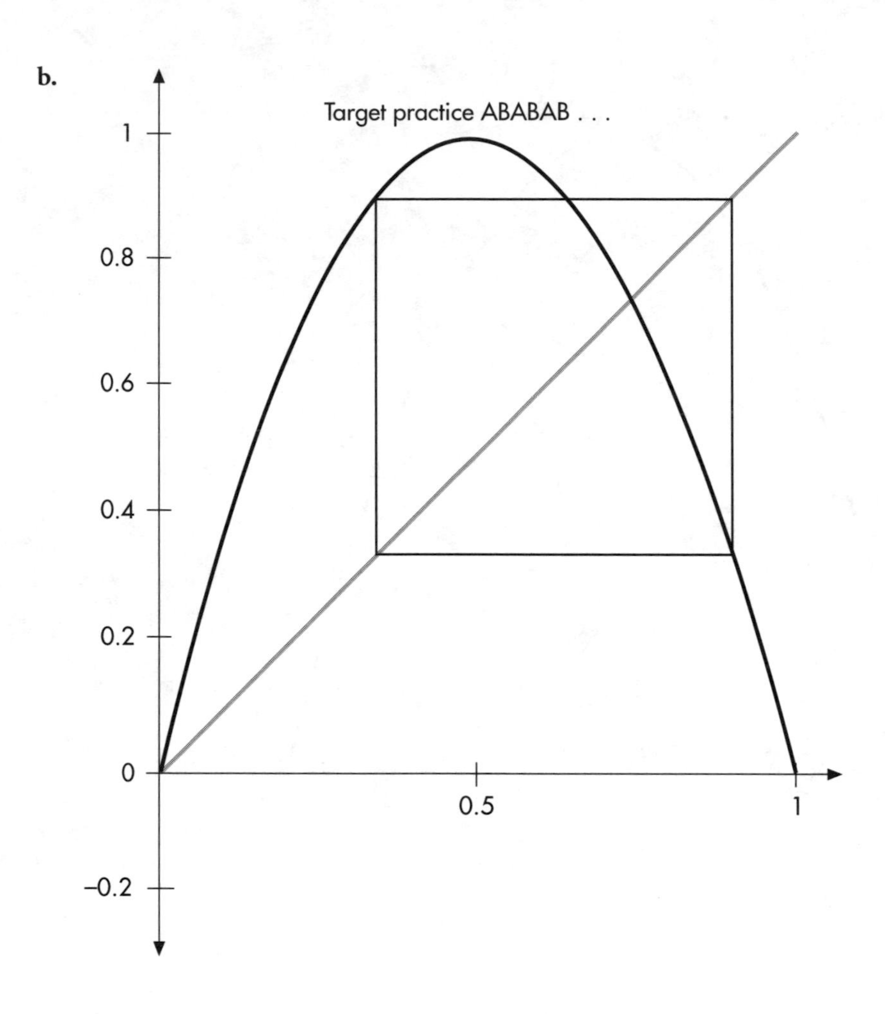

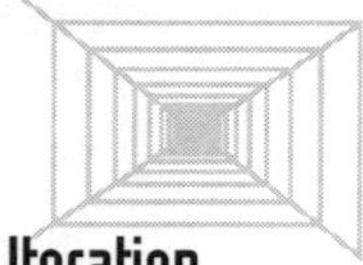

c.

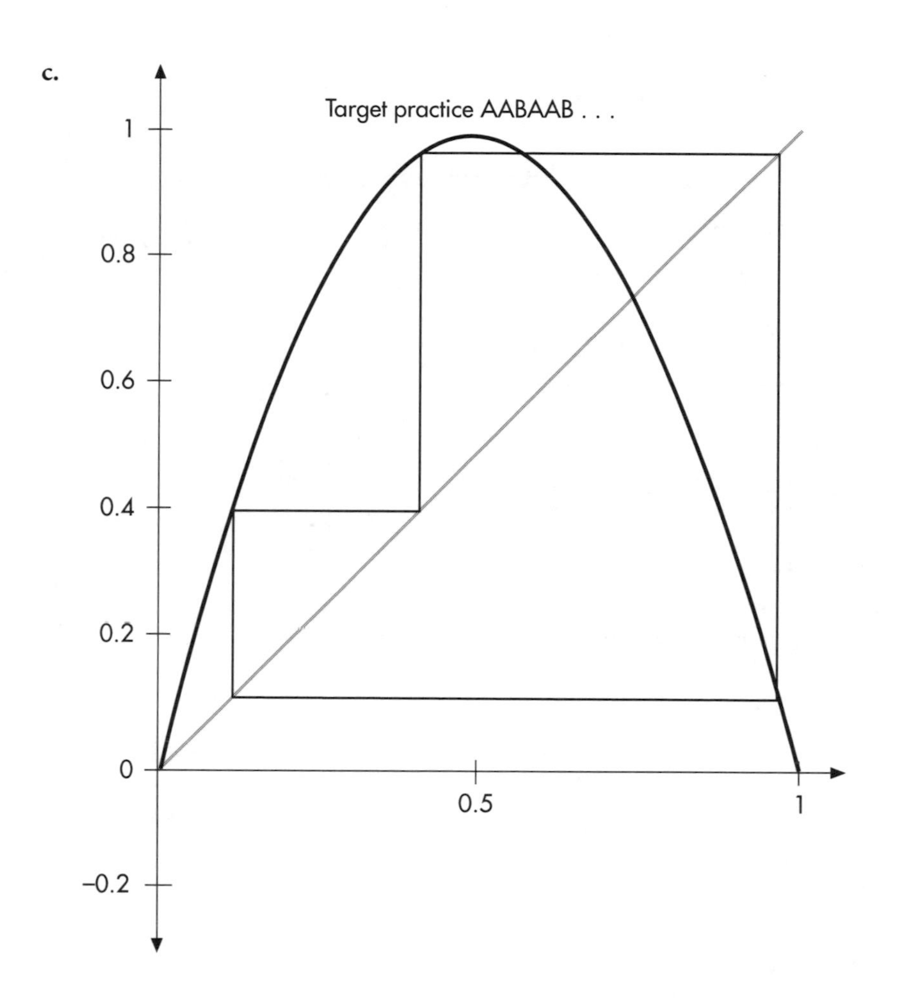

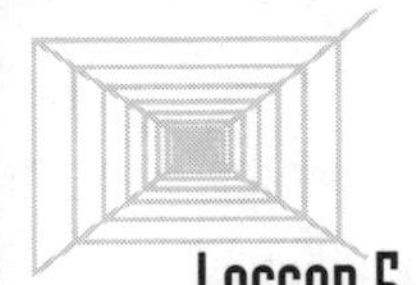

d.

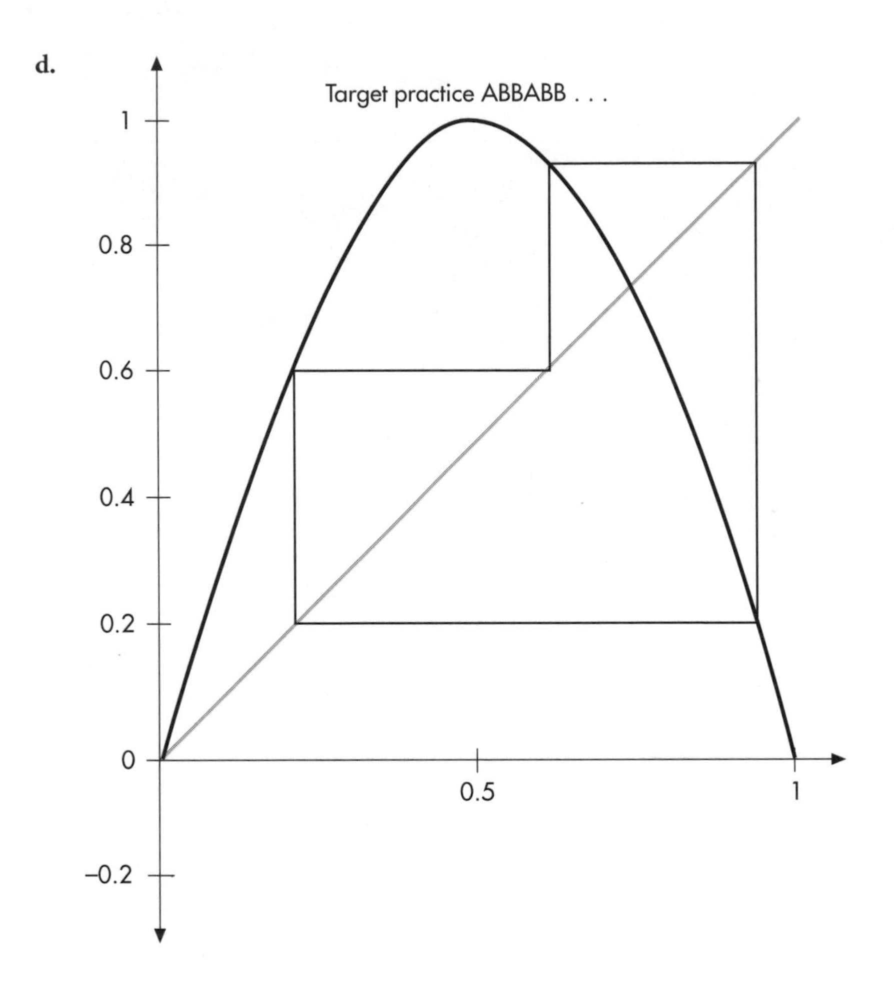

INVESTIGATION 9: A QUADRATIC ITERATION RULE

$x \to x^2 + c$ has fixed points at $x = \frac{1}{2} \pm \frac{\sqrt{1-4c}}{2}$.

a. If $1 - 4c < 0$, which implies that $c > \frac{1}{4}$, then there are no fixed points.

b. If $c = \frac{1}{4}$, there is one fixed point at $x = \frac{1}{2}$.

c. If $c < \frac{1}{4}$, there are two fixed points.

d. A bifurcation occurs at $c = \frac{1}{4}$.

e. There are always two fixed points: one on the right at $x = \frac{1}{2} + \frac{\sqrt{1-4c}}{2}$ which is always repelling, and one on the left which is attracting for $0.25 > c > -0.75$.

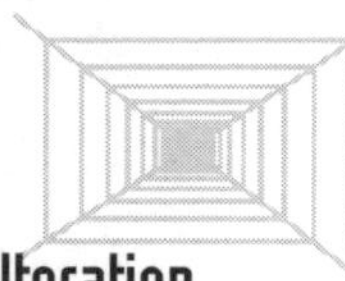

f.

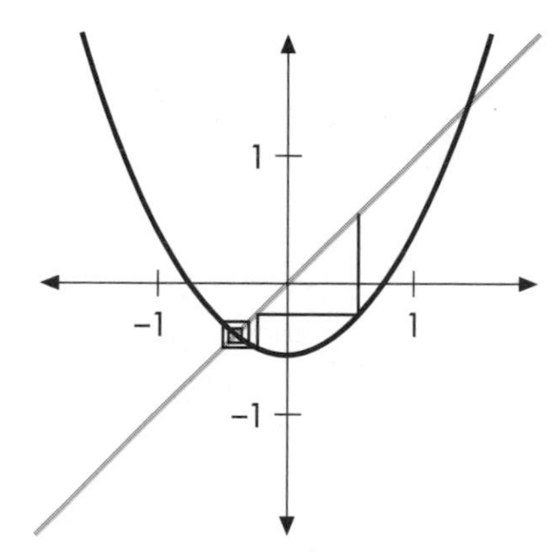

$c = -0.6$: attracting fixed point

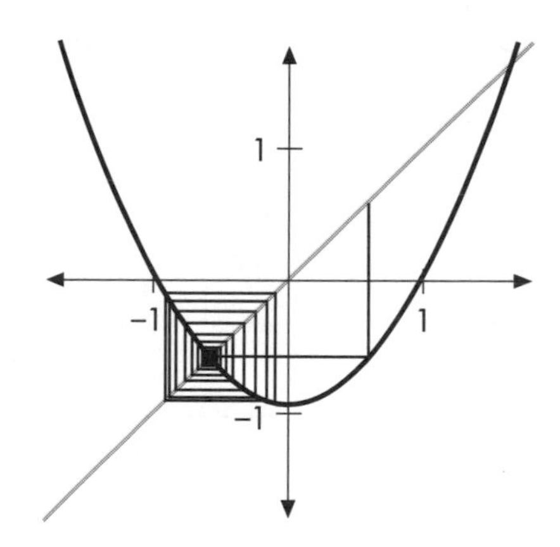

$c = -0.9$: 2-cycle

FURTHER EXPLORATION

1. **a.** There are three fixed points: -1, 0, and 1. -1 and 1 are repelling and 0 is attracting. If $x_0 < -1$, the orbits go off to negative infinity. If $x_0 > 1$, the orbits go off to positive infinity. If $-1 < x_0 < 1$, the orbits are attracted to 0.

   b. There are no fixed points and all orbits go off to infinity.

   c. There are two fixed points: 1 and -1. If x_0 is not equal to 0, all orbits become a 2-cycle of the form x_0, $^1\!/_{x_0}$, x_0, $^1\!/_{x_0}$, ...

2. There are three fixed points: *A*, *B*, and *C*. *A* and *C* are repelling and *B* is attracting.

Seed interval	**Fate of orbit**
$x_0 < A$	goes to negative infinity
$A < x_0 < B$	is attracted to *B* (or is eventually fixed at *B*)
$B < x_0 < C$	is attracted to *B* (or is eventually fixed at *B*)
$C < x_0$	goes to positive infinity

3. All orbits will be attracted to 0.73908 . . . .

4. All orbits are attracted to 0 but very, very slowly.

5. If $x_0 > 0$, all orbits go to 0. If $x_0 < 0$, then all orbits will be attracted to the fixed point of the iteration rule $x \rightarrow (\frac{1}{2})x + 1$, which is 2. Since 2 is an attracting fixed point, any orbit that starts with a seed less than 0 must eventually become positive. Once this happens, the iteration rule becomes $x \rightarrow (\frac{1}{2})x$, which has an attracting fixed point of 0 and the orbits will go to 0. Therefore, all orbits are attracted to 0.

LESSON 6 ▹ CHAOS

INVESTIGATION 1: CONSTRUCTING HISTOGRAMS

a. Here is the histogram for a seed of 0.3:

11										×
10										×
9										×
8										×
7	×	×								×
6	×	×						×		×
5	×	×						×		×
4	×	×				×	×	×	×	×
3	×	×		×	×	×	×	×	×	×
2	×	×	×	×	×	×	×	×	×	×
1	×	×	×	×	×	×	×	×	×	×
0	0.0	0.1	0.2	0.3	0.4	0.5	0.6	0.7	0.8	0.9

Here is the histogram for a seed of 0.9:

13										×
12	×									×
11	×									×
10	×									×
9	×									×
8	×									×
7	×									×
6	×									×
5	×	×		×		×				×
4	×	×		×	×	×			×	×
3	×	×	×	×	×	×			×	×
2	×	×	. ×	×	×	×			×	×
1	×	×	×	×	×	×	×	×	×	×
0	0.0	0.1	0.2	0.3	0.4	0.5	0.6	0.7	0.8	0.9

b. and **c.** In both cases, peaks are beginning to build on the extreme edges (near 0 and 1). This is more pronounced if you start with a seed of 0.9. If you look at what happens after several hundred iterations, the peaks begin to emerge on the sides. About 25% of the iterations fall between 0 and 0.1, and another 25% fall between 0.9 and 1.

INVESTIGATION 2: SIMPLER HISTOGRAMS

There is little transient behavior exhibited in these two examples. In both cases, the orbit settles into its attracting cycle quite quickly.

a. $x \rightarrow 3.25x(1 - x)$

Seed 0.25

The orbit settles into a 2-cycle (0.495 → 0.812 → 0.495 → · · ·) fairly quickly.

13										
12					×				×	
11					×				×	
10					×				×	
9					×				×	
8					×				×	
7					×				×	
6					×				×	
5					×				×	
4					×				×	
3					×				×	
2					×	×		×	×	
1			×		×	×	×	×	×	
0	0.0	0.1	0.2	0.3	0.4	0.5	0.6	0.7	0.8	0.9

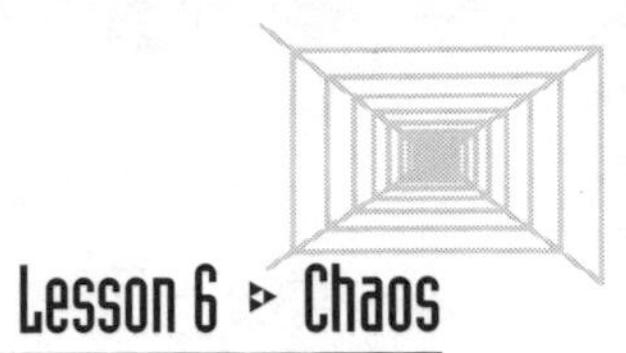

b. $x \rightarrow 3.84x(1 - x)$

Seed 0.4

The orbit settles into a 3-cycle ($0.15 \rightarrow 0.49 \rightarrow 0.96 \rightarrow 0.15 \rightarrow \cdots$) fairly quickly.

10										
9										×
8		×								×
7		×			×					×
6		×			×					×
5		×			×					×
4		×			×					×
3		×			×					×
2		×			×	×				×
1		×	×		×	×	×	×	×	×
0	0.0	0.1	0.2	0.3	0.4	0.5	0.6	0.7	0.8	0.9

Investigation 3: Time series and histograms

10										
9								×		
8								×	×	
7								×	×	
6						×		×	×	
5						×		×	×	
4			×	×		×	×	×	×	
3		×	×	×	×	×	×	×	×	
2		×	×	×	×	×	×	×	×	
1		×	×	×	×	×	×	×	×	
0	0.1	0.2	0.3	0.4	0.5	0.6	0.7	0.8	0.9	1

Investigation 4: A quadratic histogram

a.

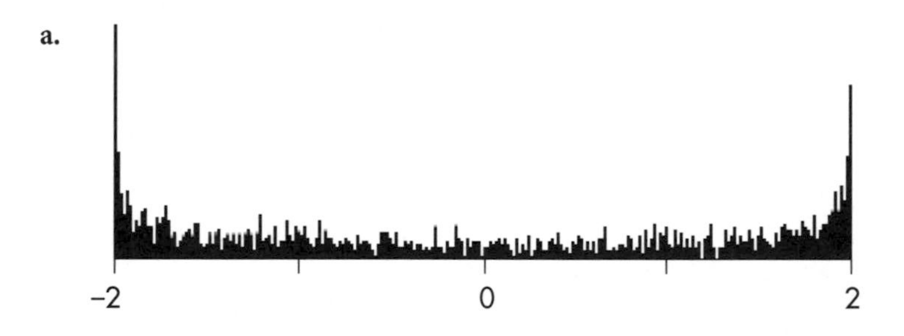

b. and c. The forward orbit of 0 is $0 \to -2 \to 2 \to 2 \to 2 \to \cdots$. The relationship between this orbit and the histogram is that there is a buildup in the histogram near -2 and 2.

INVESTIGATION 5: MATCHING HISTOGRAMS AND GRAPHICAL ITERATIONS

Histogram	Web
A	III
B	IV
C	II
D	I

FURTHER EXPLORATION

1. Both orbits settle into the same 3-cycle, but the orbit of 0.1 has a little more transient behavior. Remember, if there is an attracting cycle, the critical point (0.5 in this case) goes to that cycle and gets there quickly.

2. The behavior here is very similar to that seen in problem 1, except in this case orbits settle into a 5-cycle. Again, the orbit of 0.1 has a little more transient behavior.

LESSON 7 ▹ THE BUTTERFLY EFFECT

INVESTIGATION 1: SENSITIVITY TO INITIAL CONDITIONS

The orbits of 0.01 and 0 will differ by 0.5 after 3 iterations.

The orbits of 0 and 0.00001 will differ by 0.5 after 8 iterations.

The orbits of 0 and 0.000001 will differ by 0.5 after 10 iterations.

The orbits of 0.01 and 0.00001 will differ by 0.5 after 3 iterations.

The orbits of 0.01 and 0.000001 will differ by 0.5 after 3 iterations.

The orbits of 0.000001 and 0.00001 will differ by 0.5 after 9 iterations.

INVESTIGATION 2: GRAPHICAL ITERATION AND SENSITIVITY

The two orbits will diverge very quickly. This shows sensitive dependence on initial conditions. Two seeds that differ by a smudge in a line will produce very different orbits.

INVESTIGATION 3: ANOTHER LOGISTIC ITERATION RULE

Answers will vary here. If you started with seeds of 0.45092 and 0.45093, the orbits would differ by 0.1 by the 14th iteration and by 0.5 after the 17th iteration.

INVESTIGATION 4: MORE LOGISTIC ITERATION RULES

Only the logistic iteration rules in parts c and d show sensitivity to initial conditions. In part a, orbits tend to a 2-cycle. In part b, orbits tend to a 3-cycle.

INVESTIGATION 5: A PROJECT

Different software and different technologies give very different answers. Below are the 0th, 10th, 30th, and 100th iterations produced using an Excel spreadsheet, Prof. Devaney's Dynamics software, and a TI-83.

Iteration	0	10	30	100
Excel	0.123	0.3503319	0.07460951	0.84196978
Dynamics Lab	0.123	0.350331	0.074609	0.143591
TI-83	0.123	0.3503312	0.0746100	0.00006

INVESTIGATION 6: SENSITIVITY IN THE REAL WORLD

The weather is one of the best examples. Is global warming creating major changes in the weather? Great discussion topic.

FURTHER EXPLORATION

1. Only the iteration rules in parts a and c show sensitivity to initial conditions.

2. Try this one on a spreadsheet like Excel. The orbit of 0.2 created by Excel eventually goes to 0. However, it can be shown by a simple hand calculation that 0.2 actually is part of a 4-cycle: $0.2 \to 0.4 \to 0.8 \to 0.6 \to 0.2 \to \cdots$. On the other hand, a TI-83 will handle this calculation exactly.

   As for sensitivity, nearby orbits must diverge because each time you iterate, you are effectively multiplying by 2. Therefore, small discrepancies in the initial seed will be doubled at each iteration, forcing orbits to move apart.

LESSON 8 ▹ CYCLES AND NONLINEAR ITERATION

INVESTIGATION 1: FINDING CYCLES

Here are the time series and the graphical iterations. In plotting the graphical iterations, the first 20 iterations were omitted. Iteration rule A produces a 2-cycle; rule B, a 4-cycle; rule C, an 8-cycle; and rule D, a 3-cycle.

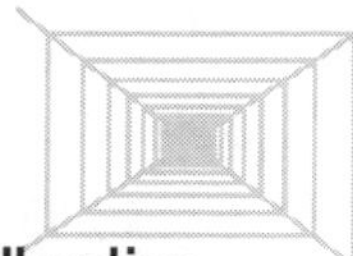

A.

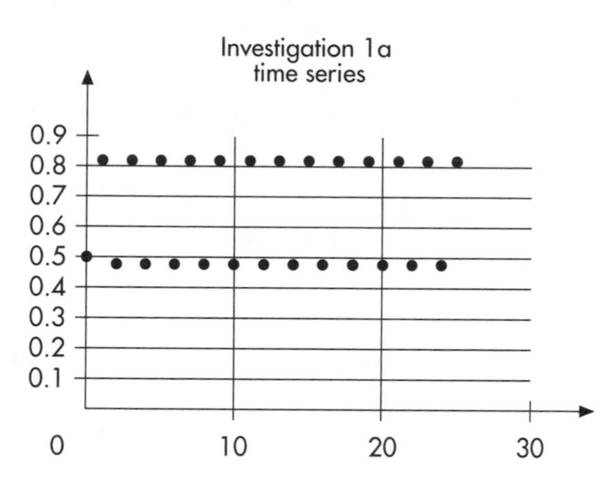
Investigation 1a
time series
0.9
0.8
0.7
0.6
0.5
0.4
0.3
0.2
0.1
0
10
20
30

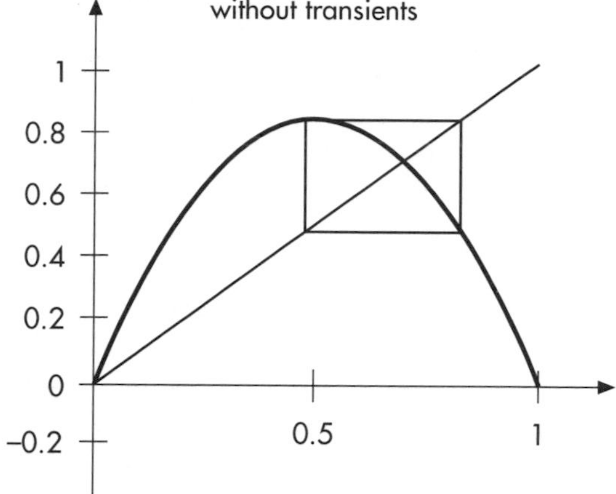
Investigation 1a orbit
without transients
1
0.8
0.6
0.4
0.2
0
−0.2
0.5
1

B.

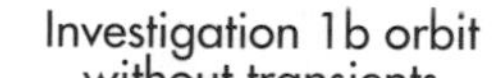
Investigation 1b orbit
without transients

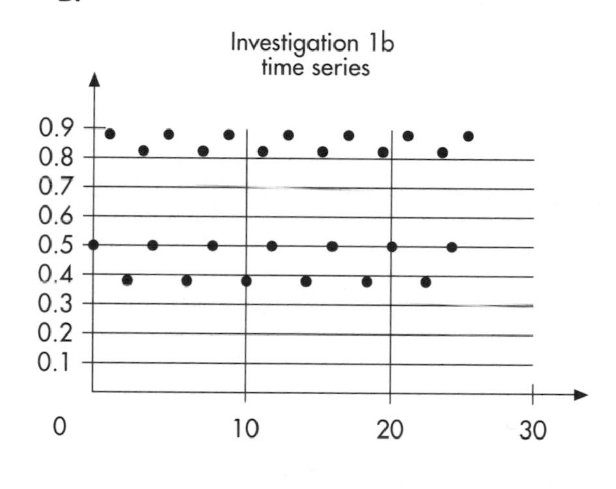
Investigation 1b
time series
0.9
0.8
0.7
0.6
0.5
0.4
0.3
0.2
0.1
0
10
20
30

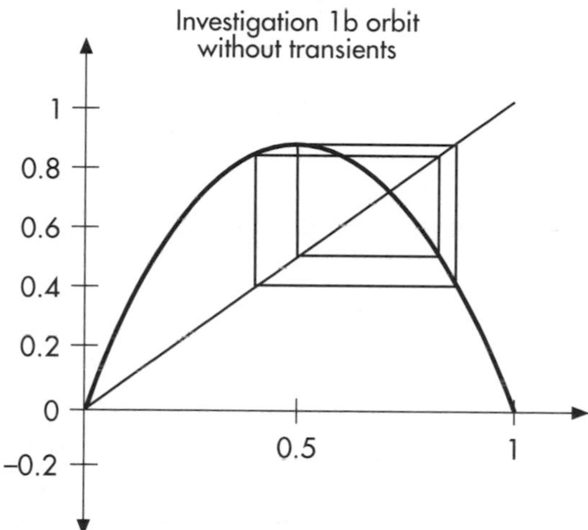
1
0.8
0.6
0.4
0.2
0
−0.2
0.5
1

C.

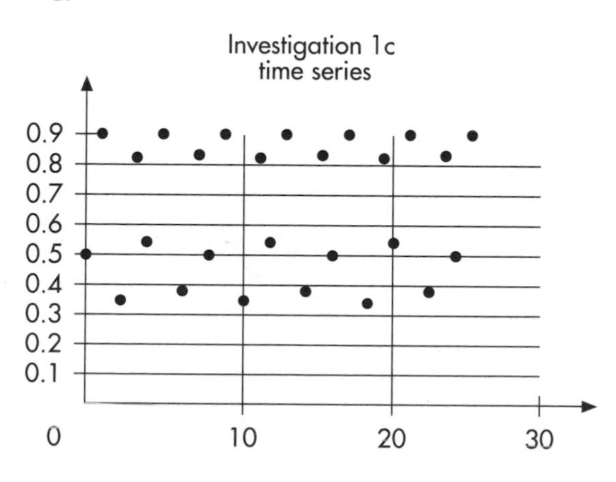
Investigation 1c
time series
0.9
0.8
0.7
0.6
0.5
0.4
0.3
0.2
0.1
0
10
20
30

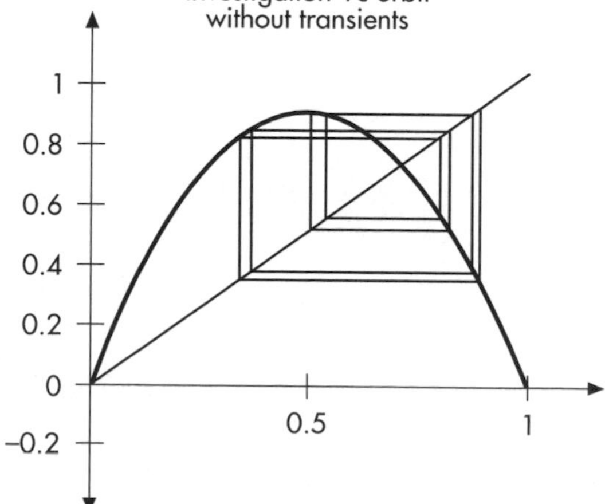
Investigation 1c orbit
without transients
1
0.8
0.6
0.4
0.2
0
−0.2
0.5
1

D.

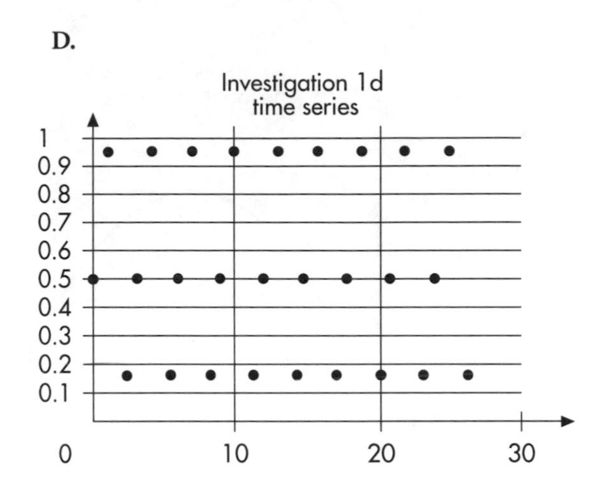

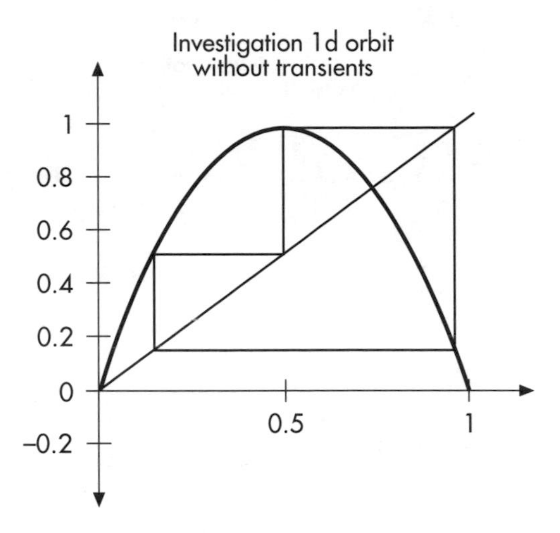

INVESTIGATION 2: FIXED POINTS AND 2-CYCLES

Answers will vary. A possible response is

Nature of orbit	Values of k
Tends to fixed point	2.8, 2.85, 2.9
Tends to 2-cycle	3.05, 3.1, 3.15, 3.2
Cannot determine	2.95, 3, 3.01

INVESTIGATION 3: FINDING OTHER CYCLES

Using a k-value of 3.4 produces a 2-cycle. k-values of 3.41, 3.42, 3.43, and 3.44 produce orbits that take a long time to settle down. Values of 3.45, 3.46, 3.47, . . . , 3.5 produce 4-cycles.

INVESTIGATION 4: FINDING CYCLES GRAPHICALLY

As k takes on values from 2.8 to 3.2, the graph of the second iteration goes from intersecting $y = x$ in two points to intersecting in four points. The two new points are fixed for the second iteration and hence form a 2-cycle as explained in the text. The transition takes place at $k = 3$.

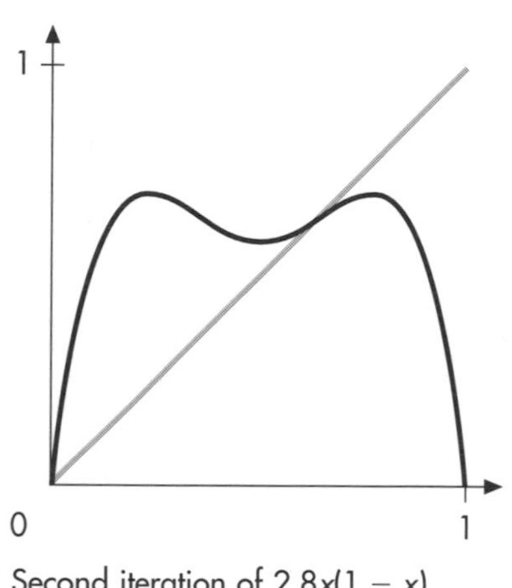

Second iteration of $2.8x(1 - x)$

Second iteration of $3.2x(1 - x)$

INVESTIGATION 5: HIGHER ITERATIONS

a. The fourth iteration crosses the diagonal 16 times and the fifth iteration 32 times.

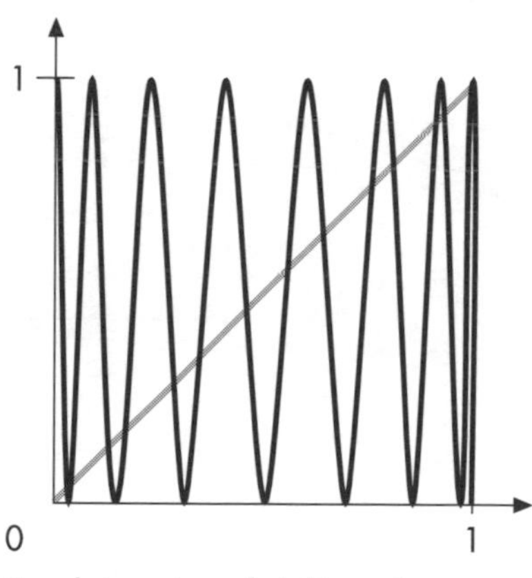

Fourth iteration of $4x(1 - x)$

b. The nth iteration will cross the diagonal 2^n times.

INVESTIGATION 6: FINDING A 3-CYCLE

This investigation can be approached analytically. If you have access to a graphing calculator that has an intersect feature, you can find the eight points of intersection of the third iteration of $x \rightarrow 4x(1 - x)$ with the diagonal. If you were to use a TI-83, you would get the following points:

0, 0.1169778, 0.1882551, 0.41317591, 0.61126047, 0.75, 0.95048443, 0.96984631

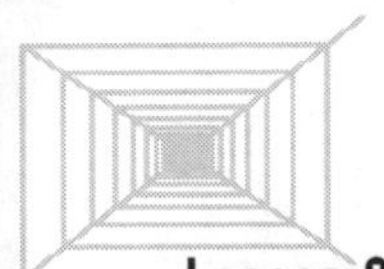

0 and 0.75 are fixed points, but the others lie on 3-cycles.

Here is the orbit beginning at 0.1882551:

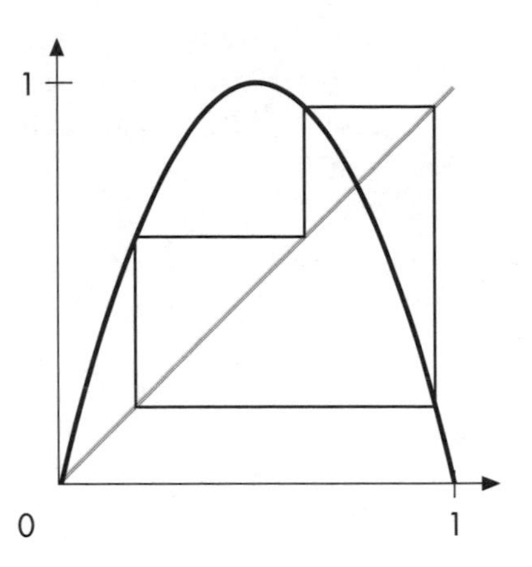

INVESTIGATION 7: FINDING A 4-CYCLE

There are 16 points of intersection of the graph of the fourth iteration with the diagonal. Two are fixed points, two lie on a 2-cycle, and the rest lie on 4-cycles.

The graph below was produced using a seed of 0.45386582. There are other possible 4-cycles (2 others, in fact).

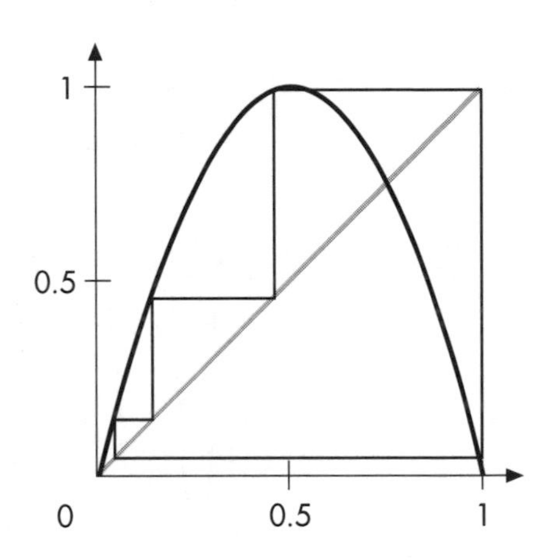

FURTHER EXPLORATION

1. The first iteration crosses the diagonal twice, yielding two fixed points:

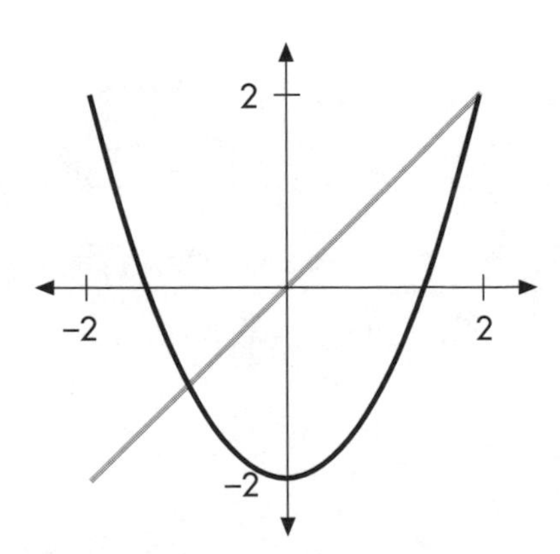

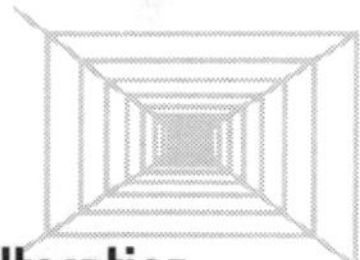

The second iteration crosses the diagonal four times, yielding two fixed points and two points on a 2-cycle:

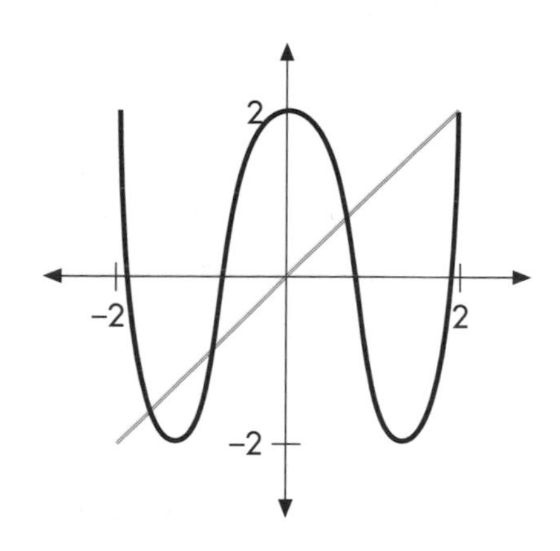

The third iteration crosses the diagonal eight times, yielding two fixed points and all other points lying on 3-cycles:

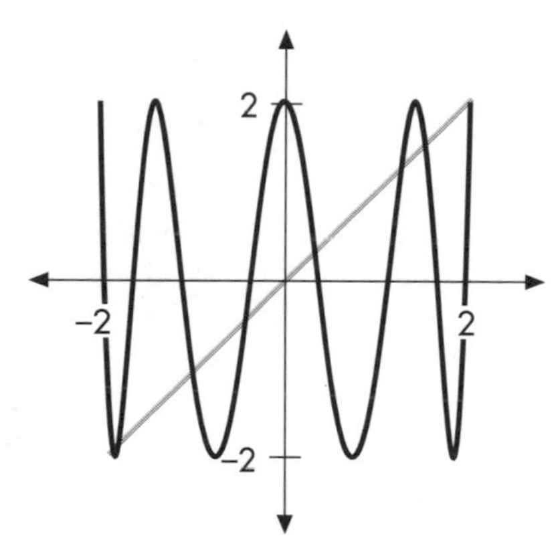

2. If there is a cycle of period 2, say $a, b, a, b, a, b, \ldots$, then a and b must be fixed for $f(f(x))$. In this case, since $f(x) = x^2 - 2$, $f(f(x)) = (x^2 - 2)^2 - 2 = x^4 - 4x^2 + 2$. Since you know that -1 and 2 are fixed points for $f(x)$, they must also be fixed for $f(f(x))$. To find the fixed points for $f(f(x))$, you have to solve the equation $x = f(f(x)) = x^4 - 4x^2 + 2$. This simplifies to solving $x^4 - 4x^2 - x + 2 = 0$. If you let $p(x) = x^4 - 4x^2 - x + 2$, then you know that if $p(a) = 0$ then $x - a$ is a factor of $p(x)$. Since you know that -1 and 2 are fixed points for $f(x)$, they must also be fixed for $f(f(x))$. But this means that $p(2) = 0$ and $p(-1) = 0$ and hence $x - 2$ and $x + 1$ are factors of $p(x)$. Dividing $p(x)$ by $(x - 2)(x + 1)$ yields

$$p(x) = x^4 - 4x^2 - x + 2 = (x - 2)(x + 1)\left(x^2 + x - 1\right)$$

To find all the roots of $p(x) = 0$, you need to solve $x^2 + x - 1 = 0$, which has solutions $x = -\frac{1}{2} \pm \frac{\sqrt{5}}{2}$, so the points of period 2 are approximately $x = 0.61803399$ and $x = -1.61803399$.

3. In this rule, if $|x| > 1$, then the orbit goes off to infinity (both plus and minus). If $|x| = 1$, then there is a 2-cycle so 1 and -1 lie on a 2-cycle. If $|x| < 1$, then the orbits all go to 0.

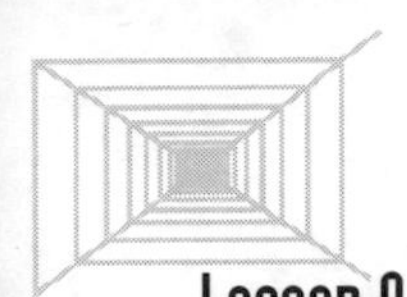

4. **a.** This turns out to be much easier if you use fractions to calculate the orbit. For example, a seed of $\frac{1}{3}$ produces the orbit

$$\frac{1}{3} \to \frac{2}{3} \to \frac{1}{3} \to \frac{2}{3} \to \cdots$$

which is a 2-cycle. A seed of $\frac{1}{9}$ produces the orbit

$$\frac{1}{9} \to \frac{2}{9} \to \frac{4}{9} \to \frac{8}{9} \to \frac{7}{9} \to \frac{5}{9} \to \frac{1}{9} \to \cdots$$

which is a 6-cycle. You can actually produce a cycle of any period k by noting that a seed of $\frac{1}{3}$ produced a cycle of period 2, a seed of $\frac{1}{7}$ will produce a cycle of period 3, a seed of $\frac{1}{15}$ will produce a cycle of period 4, and in general, a seed of $1/(2^k - 1)$ will produce a cycle of period k. To prove why this works, observe that in the sequence $1, 2, 4, 8, \ldots, 2^k$ there are always k terms less than 2^k. So in particular, $1/(2^3 - 1) = \frac{1}{7}$ produces a 3-cycle. Note that $\frac{3}{7}$ produces a different 3-cycle.

b. $\frac{1}{15}$, $\frac{3}{15} = \frac{1}{5}$, and $\frac{7}{15}$ all yield 4-cycles.

The graph of the second iteration, shown at right, crosses the diagonal in four places.

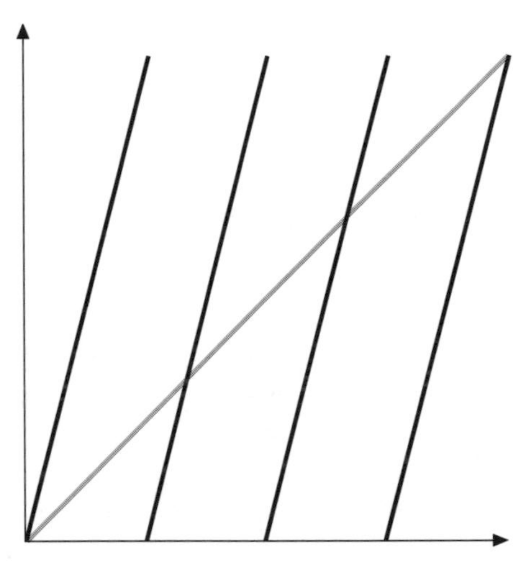

c. The graph of the third iteration, shown at right, crosses the diagonal in eight places.

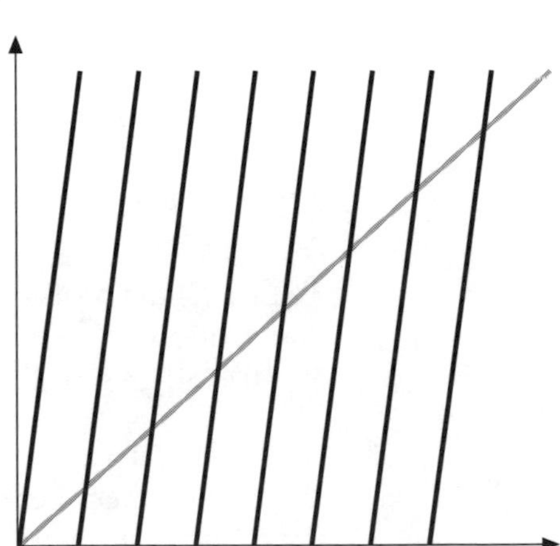

LESSON 9 ▹ THE ORBIT DIAGRAM

INVESTIGATION 1: LARGER k-VALUES

The orbit explodes. To see this, let $k = 4.1$ and calculate the orbit of 0.5. The second iteration is -0.1050625, which is less than 0. Since we have shown that 0 is a repelling fixed point, the orbit will take off to negative infinity.

Note that once an orbit is negative, it remains negative and goes off to negative infinity.

INVESTIGATION 2: FINDING WINDOWS

Here are some of the many possible windows:

Period 4: 3.960 to 3.961. This is one of the harder windows to find! Note that the period 4 region after the original period-doubling is *not* a period 4 window—it is part of the period 1 window.

Period 5: 3.738 to 3.744
Period 6: 3.626 to 3.634
Period 7: Near 3.774 . . . Also near 3.701 . . .
Period 8: Near 3.662 . . .
Period 9: Near 3.6872 . . .
Period 10: Near 3.647

INVESTIGATION 3: WORLD'S LONGEST ORBIT DIAGRAM

Answers will vary.

INVESTIGATION 4: WINDOWS BETWEEN WINDOWS

a. The original two windows have periods 6 and 5:

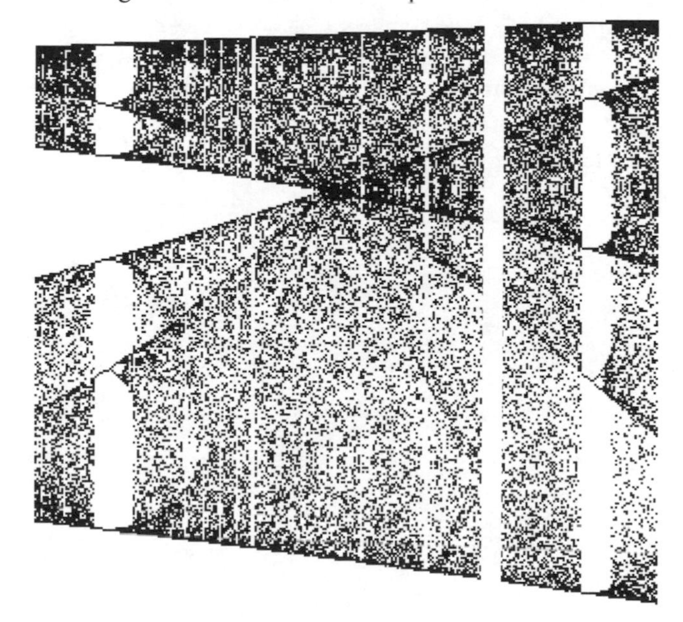

b. The largest windows between period 1 and period 6 (the leftmost window in part a) have periods 12 and 10:

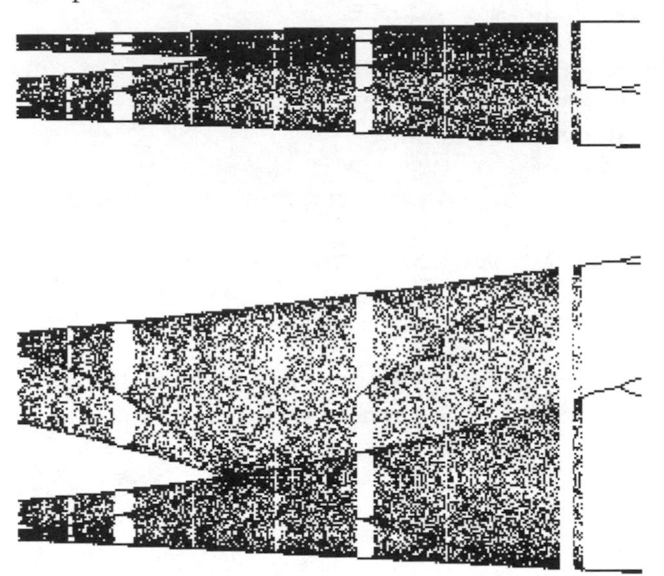

c. At the next stage, the periods are 24 and 20, and they continue to double in period as you move closer to the period 1 window.

Investigation 5: Smears in the Orbit Diagram

The smears occur at the bifurcation points because the graph is about to split and it takes a large number of iterations to settle into its cycle. By taking larger and larger numbers of iterations, the smears can be smoothed out, but never completely, due to roundoff error in the computer.

Further Exploration

1. The two orbit diagrams are very similar in many respects. They both exhibit period-doubling, and they both have the surprise period-3 window. The two orbit diagrams are shown on the following page.

Here is the orbit diagram for $f(x) = k(x - (x^3/3))$ on the interval $1.961 \leq k \leq 2.598$:

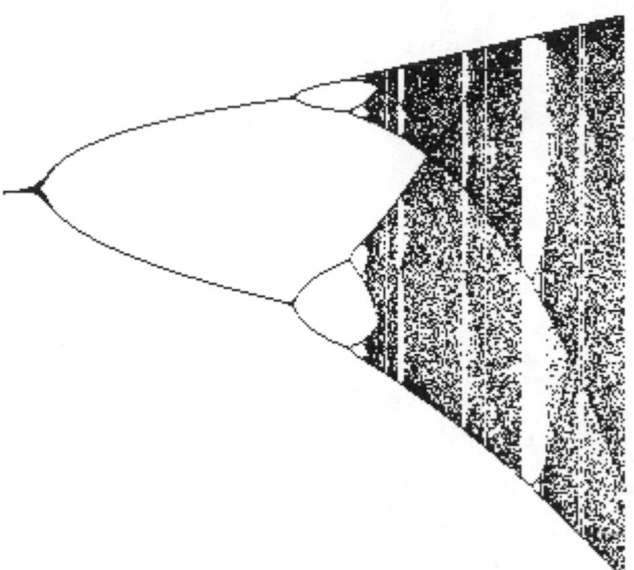

Here is the orbit diagram for the function $f(x) = kx(1 - x)$ on the interval $2.85 < k < 4.0$:

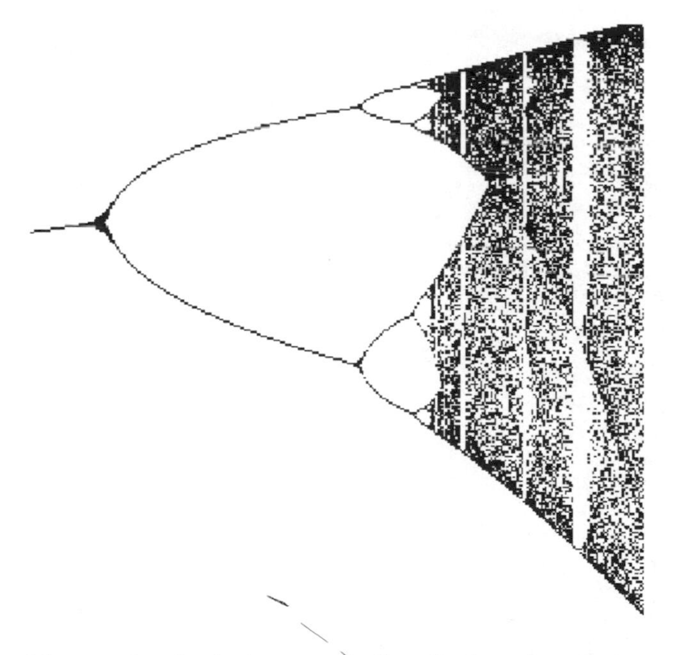

2. To see why the hole appears, first find a value C of c such that the second iteration of $1/2$ is equal to the fixed point. Second, show that for values of c such that $1 < c < C$, the second iteration of $1/2$ is just slightly larger than the fixed point. Since the fixed point is repelling, this leaves a gap, hence the hole.

0 is a fixed point for this function, which goes from being attracting to repelling at $c = 1$. Also, for c slightly greater than 1 another fixed point is created (since $y = x$ now intersects $y = c(1 - x)$), which is also repelling, so you begin to get the gap mentioned in the preceding paragraph.

3. Here is the graph of $2.8 \cos(x)$:

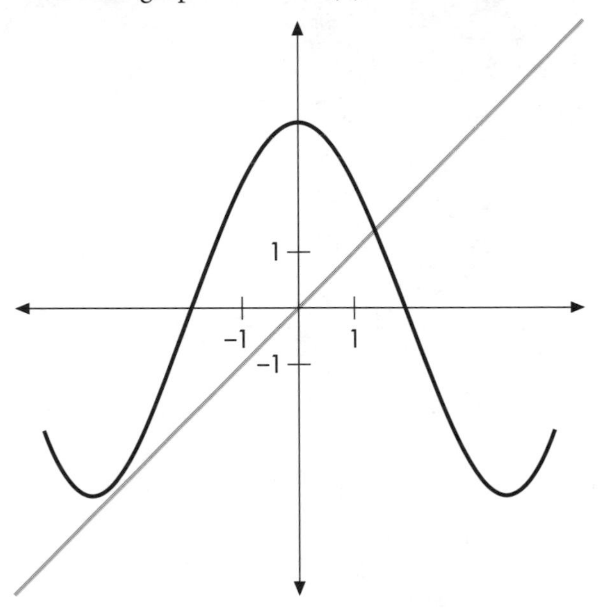

Here is the graph of $3 \cos(x)$:

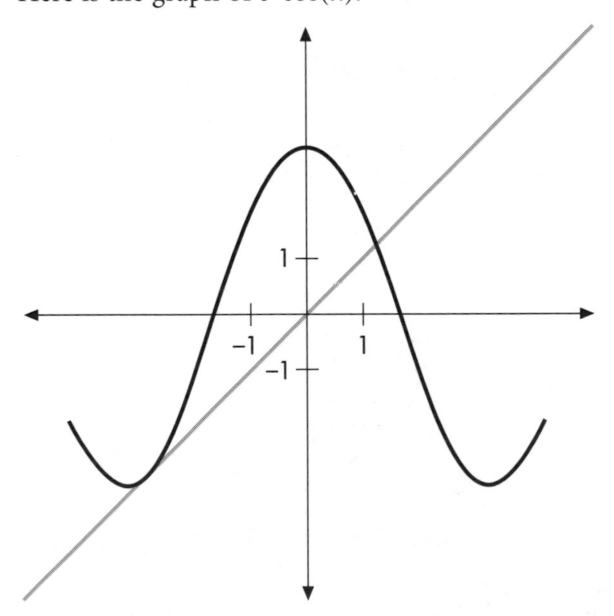

Note how a tangency has occurred. If we start with the seed 0, in the second case the orbit is trapped by an attracting fixed point:

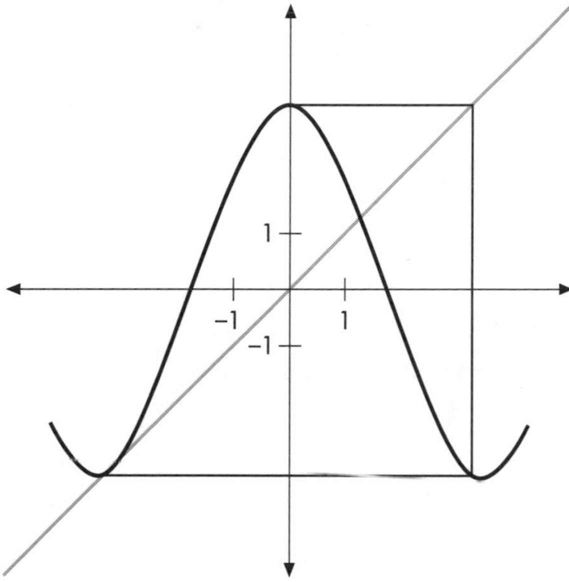

But in the first case, there is no attracting fixed point, and so the orbit runs around the interval:

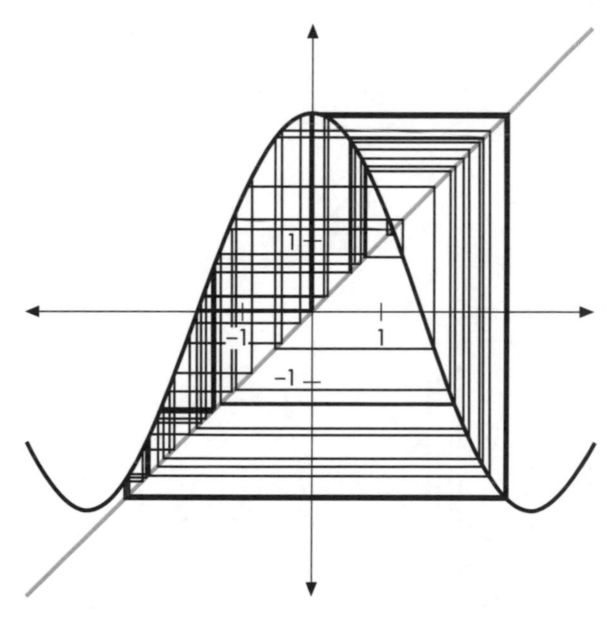

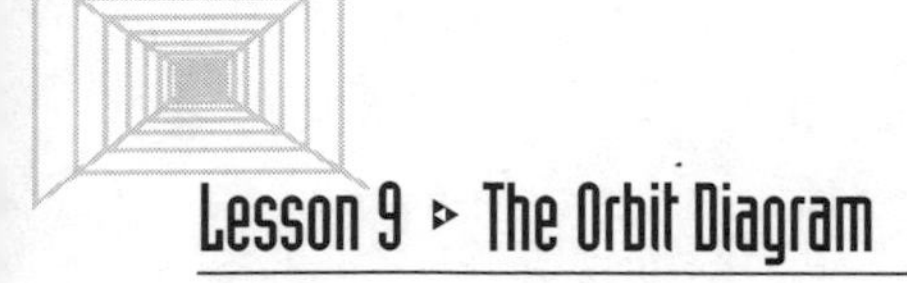

The orbits remain trapped in a tiny interval for A between 2.975 . . . and 4.19 as shown here for $A = 4$:

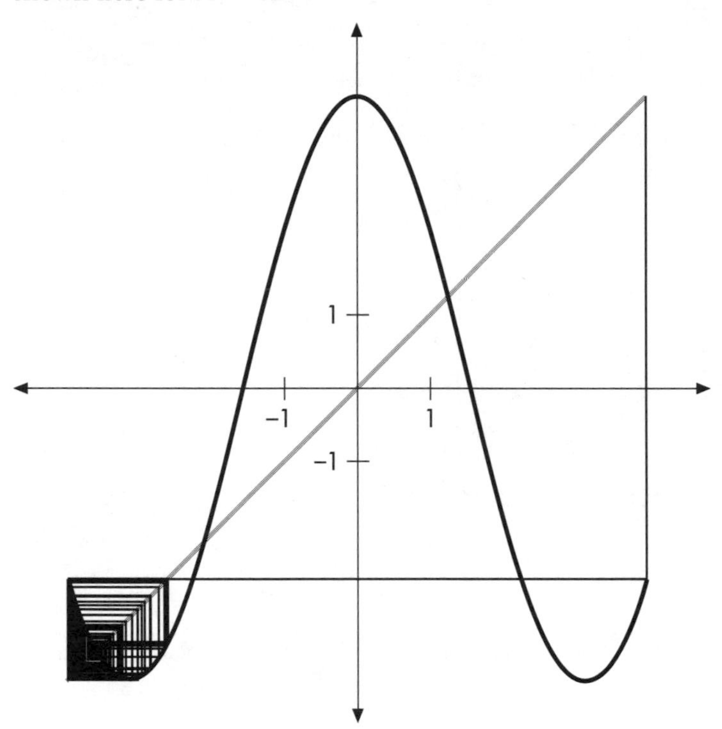

When A is larger than 4.19 . . . , this orbit can escape from the interval:

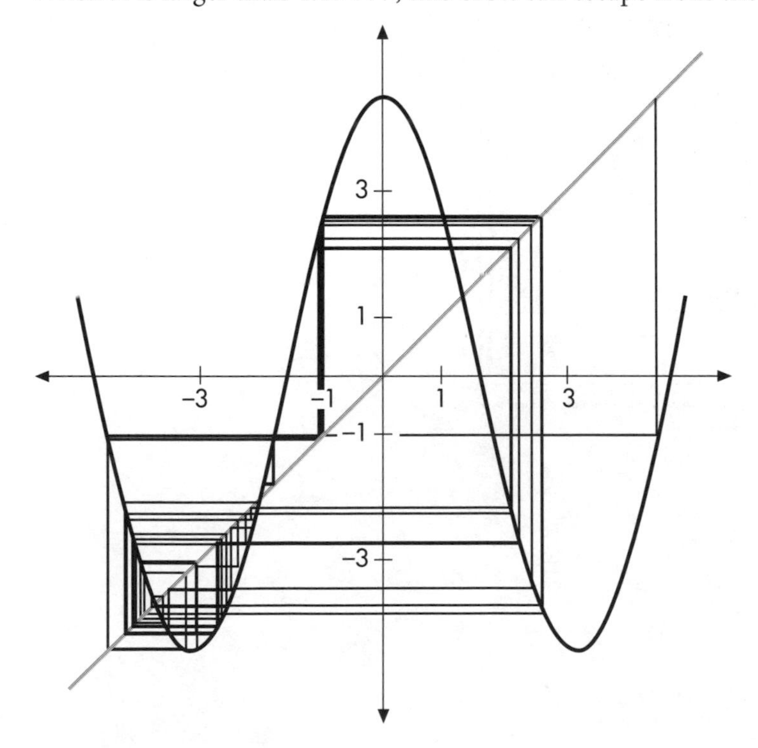

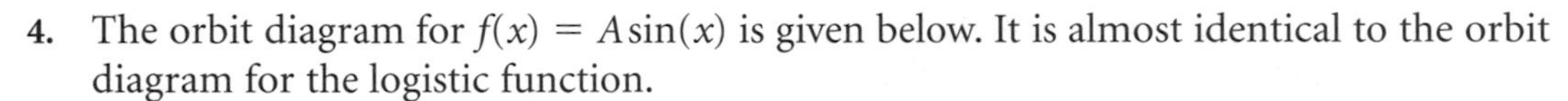

4. The orbit diagram for $f(x) = A\sin(x)$ is given below. It is almost identical to the orbit diagram for the logistic function.

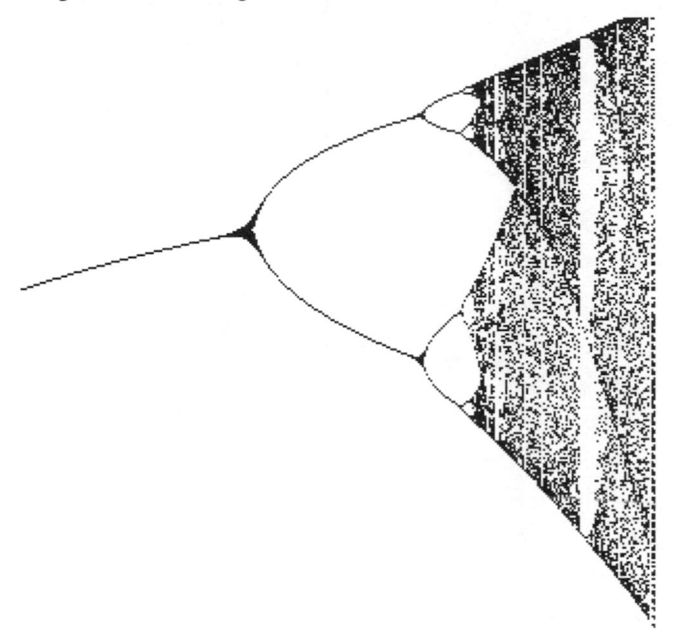

LESSON 10 ▹ A QUADRATIC EXPEDITION

INVESTIGATION 1: NON-ESCAPE *c*-VALUES

a. The orbit of 0 will not go to infinity if $-2 \le c \le 0.25$. If $c > 0.25$, then there are no fixed points. If $c < -2$, then the orbit of 0 goes off to infinity.

b. Here is one way to show this. First, show that if $c < 1/4$ then $x^2 + c$ has two fixed points: one at $x = 1 + \sqrt{1 - 4c}/2$ and the other at $x = 1 - \sqrt{1 - 4c}/2$. For the orbit of 0 to be bounded, the second iteration, $c^2 + c$, must be less than $1 + \sqrt{1 - 4c}/2$ since it has already been shown that $1 + \sqrt{1 - 4c}/2$ is a repelling fixed point. The inequality $c^2 + c > 1 + \sqrt{1 - 4c}/2$ has as a solution $c < -2$, and hence all orbits of 0 produced by iterating $x^2 + c$ will go off to infinity if $c < -2$.

Graphical iteration shows the solutions and the orbit going to infinity when $c = -2.1$:

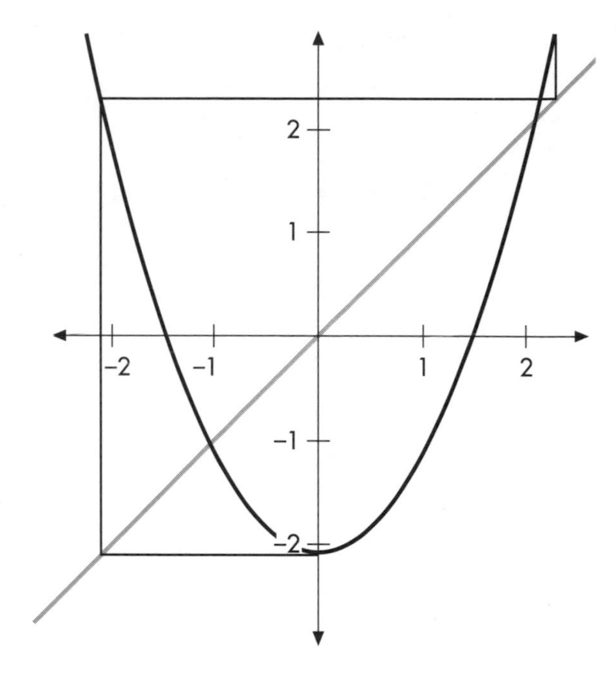

INVESTIGATION 2: FIXED POINTS

As shown in the preceding Investigation, $x^2 + c$ has two fixed points: one at $x = 1 + \sqrt{1 - 4c}/2$ and the other at $x = 1 - \sqrt{1 - 4c}/2$. Since the fixed points are defined if and only if $1 - 4c \geq 0$, c must be less than or equal to $1/4$. There are no fixed points if $c > 1/4$. There is one fixed point if $c = 1/4$, and there are two fixed points if $c < 1/4$.

INVESTIGATION 3: ATTRACTING FIXED POINTS

The iteration rule $x \to x^2 + c$ has an attracting fixed point for $-0.75 \leq c \leq 0.25$. One way to see this is that for $c < -0.75$ there is an attracting 2-cycle. For c slightly greater than -0.75, there is an attracting fixed point. As c goes from being slightly greater than -0.75 to slightly less than -0.75, the attracting fixed point becomes repelling; meanwhile, an attracting 2-cycle is born.

INVESTIGATION 4: OTHER ORBITS

a. All orbits go off to infinity.

b. If $|x_0| < 1$, the orbits go to 0.
If $x_0 = 1$, the orbit is fixed.
If $x_0 = -1$, the orbit is fixed after one iteration.
If $|x_0| > 1$, the orbits go off to infinity.

c. This rule has a repelling fixed point at $(1 + \sqrt{3})/2$ and an attracting one at $(1 - \sqrt{3})/2$. If $x_0 = -(1 + \sqrt{3})/2$, the orbit is fixed. If $|x_0| < (1 + \sqrt{3})/2$, the orbit is attracted to $(1 - \sqrt{3})/2$. If $|x_0| > (1 + \sqrt{3})/2$, the orbits go off to infinity.

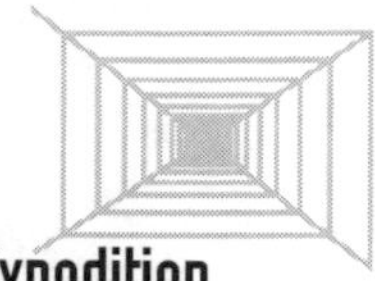

d. This rule has two repelling fixed points at $1 + \sqrt{5}/2$ and $1 - \sqrt{5}/2$. If $x_0 = -1 + \sqrt{5}/2$, the orbit is eventually fixed. If $|x_0| < 1 + \sqrt{5}/2$, the orbits are attracted to a 2-cycle at 0 and -1. If $|x_0| > 1 + \sqrt{5}/2$, the orbits go off to infinity.

INVESTIGATION 5: NON-ESCAPE ORBITS

$x \to x^2 + c$ will always have two fixed points if $c < 0.25$: one at $x = 1 + \sqrt{1-4c}/2$ and the other at $x = 1 - \sqrt{1-4c}/2$. The one at $x = 1 + \sqrt{1-4c}/2$ will always be repelling. Since the graph of $y = x^2 + c$ is symmetric with respect to the y-axis, if $x_0 < -1 + \sqrt{1-4c}/2$, the first iteration of x_0 will be greater than $1 + \sqrt{1-4c}/2$ and will produce an orbit that goes off to infinity.

INVESTIGATION 6: WHAT HAS CHANGED?

a. Sketch graphs for the following values of c: -0.73, -0.75, and -0.77. You will see that something happens when $c = -0.75$.

b.–d. The iteration rule for the second iteration is $x \to (x^2 + c)^2 + c = x^4 + 2cx^2 + c^2 + c$. Here is the graph of the second iteration of $x \to x^2 - 0.4$. Notice that it intersects $y = x$ in only one place on the part of the graph shown.

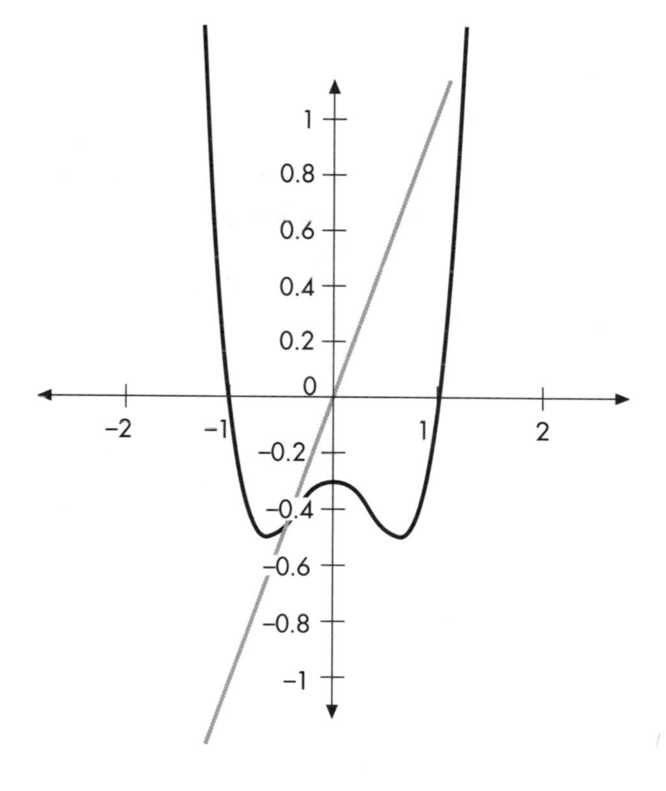

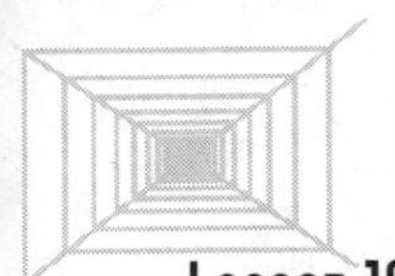

Here is the graph of the second iteration of $x \to x^2 - 0.75$. Notice that it appears to intersect $y = x$ in more than one place on the part of the graph shown.

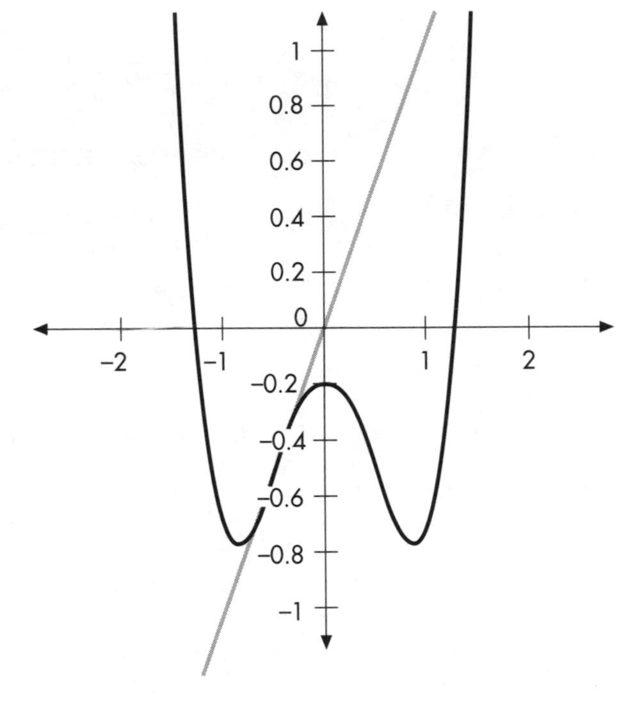

Here is the graph of the second iteration of $x \to x^2 - 0.90$. Notice that it appears to intersect $y = x$ in three places on the part of the graph shown.

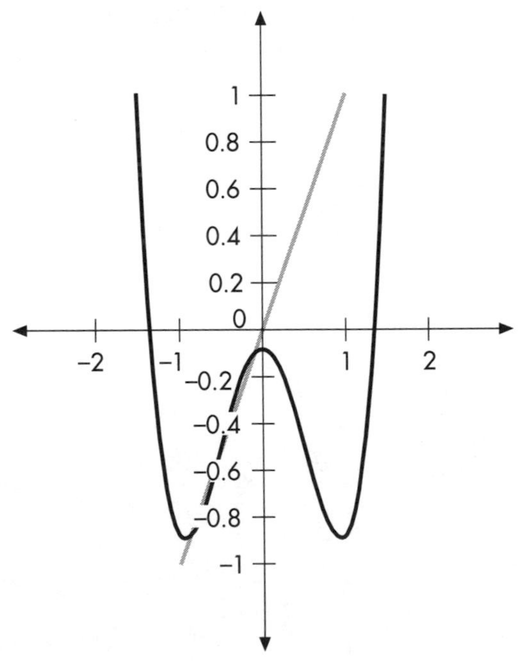

e. As you lower the c-value, the second iteration goes from having two points of intersection with the diagonal to having four. This change occurs at $c = -0.75$.

Investigation 7: Other c-values

a. $c = -1.3$: 4-cycle
$c = -1.35$: 4-cycle
$c = -1.4$: 32-cycle (this is very hard to detect!)
$c = -1.5$: chaotic
$c = -1.77$: 6-cycle
$c = -1.8$: chaotic
$c = -2$: eventually fixed

b. Here is a picture of the full orbit diagram:

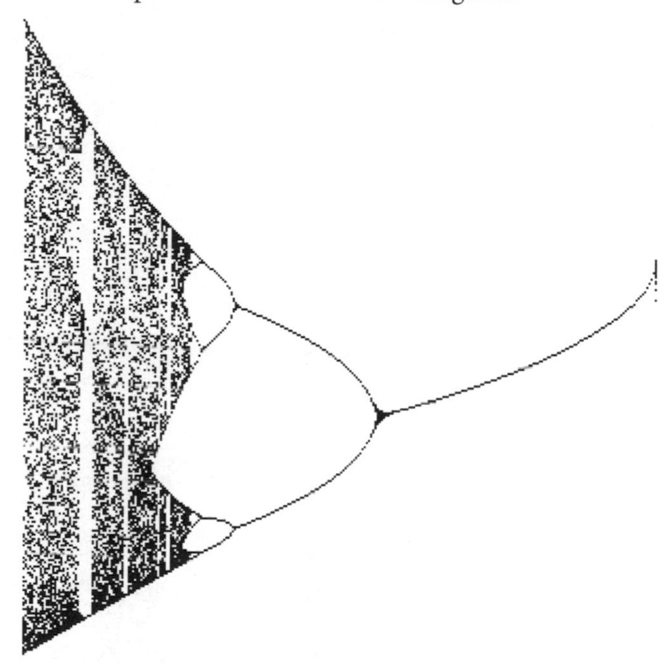

Investigation 8: A special case

a. The fixed points are 2 and -1.

b. 0 is eventually fixed at 2.

c. If $|x_0| < 2$, the orbits are bounded and are generally chaotic.

If $|x_0| = 2$, the orbit is fixed or eventually fixed.

If $|x_0| > 2$, the orbits go off to infinity.

d. There is sensitivity to initial conditions, for example, near 2.

e. Here is a time-series graph, a histogram, and a graphical iteration using a seed of 0.171942 (Jon Choate's birthday).

A time series for a seed of 0.171942:

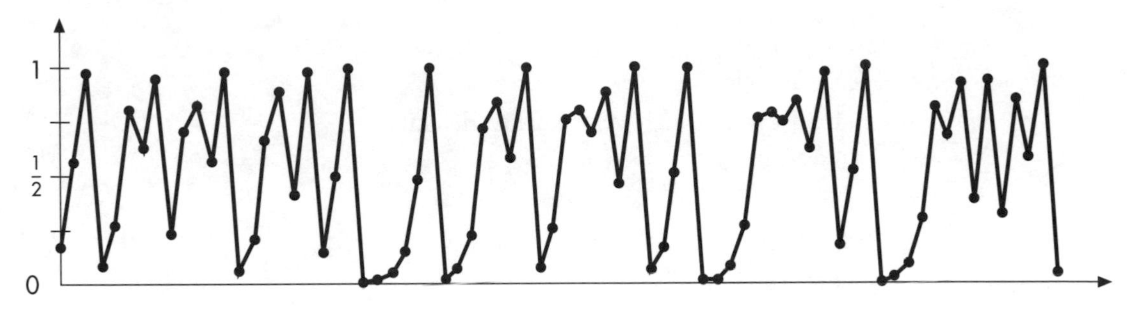

A histogram for a seed of 0.171942:

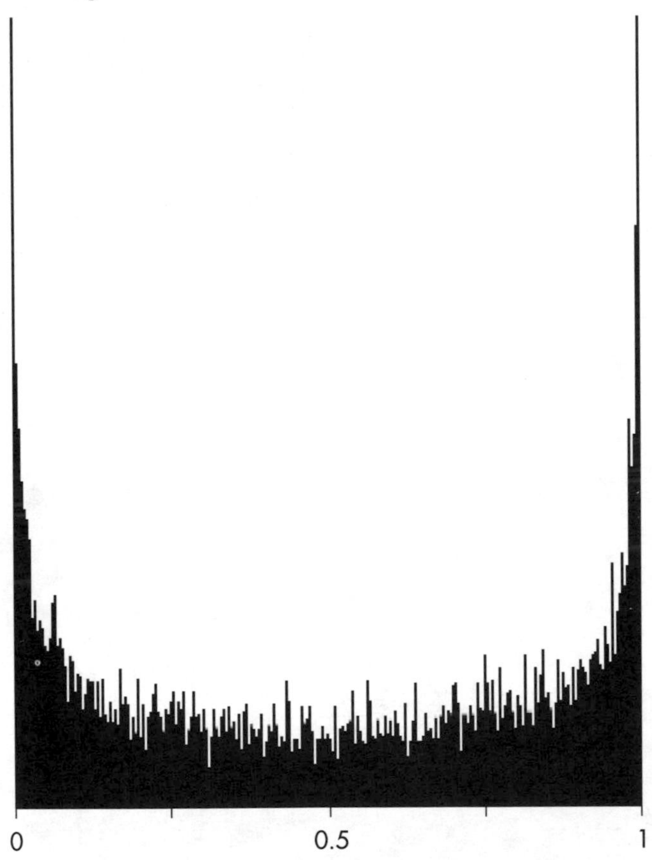

Graphical iteration for a seed of 0.171942:

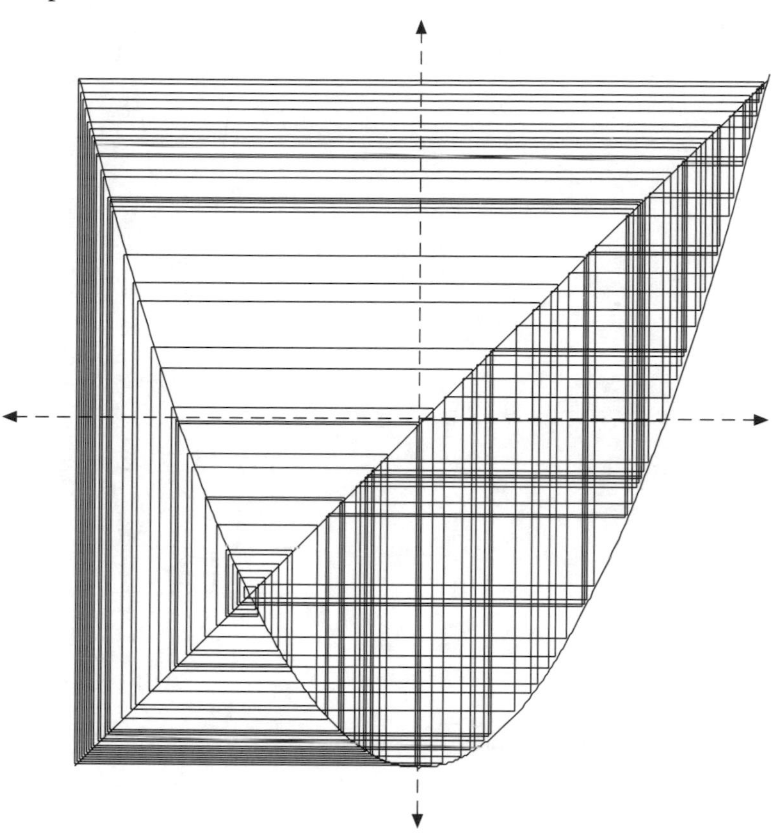

Investigation 9: Graphing higher iterations

a. The first iteration has two intersections with the diagonal:

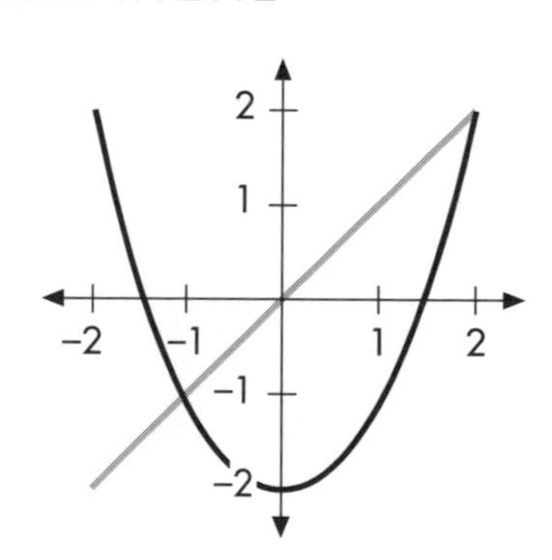

The second iteration has four intersections with the diagonal:

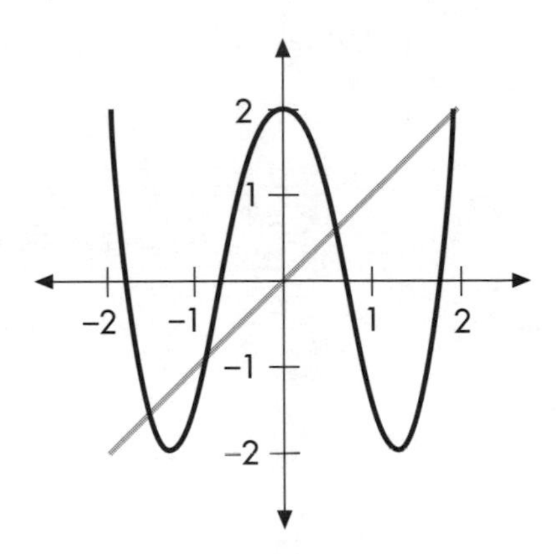

The third iteration has eight intersections with the diagonal:

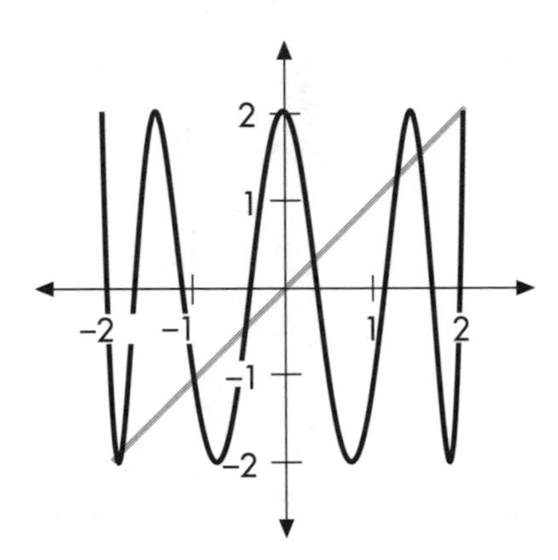

The fourth iteration has 16 intersections with the diagonal:

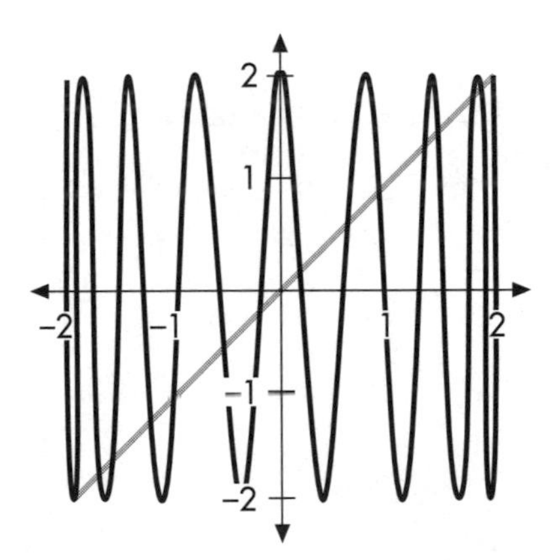

b. The *n*th iteration will have 2^n intersections with the diagonal.

Investigation 10: Target practice

To create a 3-cycle, try a seed of −1.8:

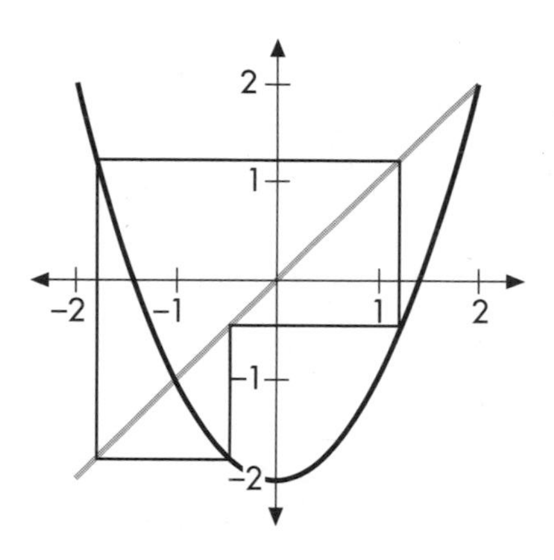

To create a 4-cycle, try a seed of −1.205:

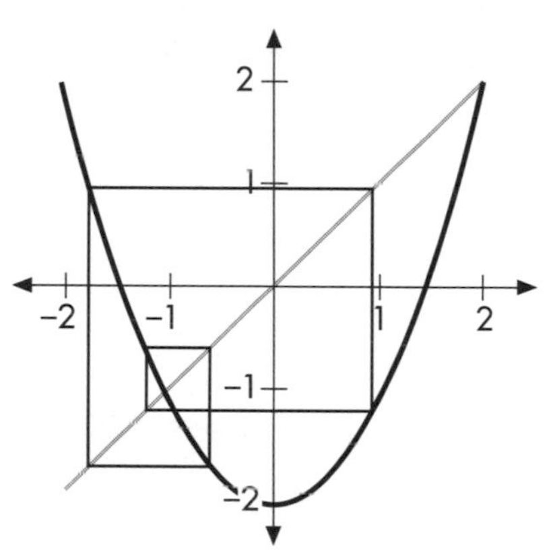

To create a 5-cycle, try a seed of −1.518:

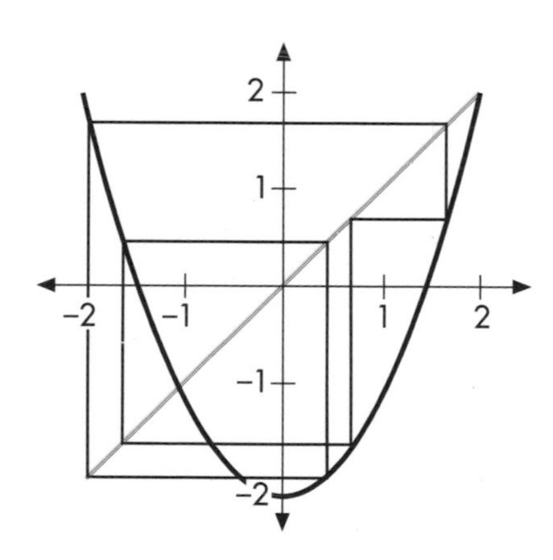